The Noise *free* Home

The four-step soundproofing method to
bring peace and quiet back to your life

JIM PRIOR

R^e think

*To my (not so quiet) children Jessica and Oliver.
See kiddies, if you set your mind to it you can
achieve anything.*

Contents

Introduction

Is your home a haven of tranquillity and calm? Or is outside noise from industry, busy roads, railways or air traffic polluting your peace and quiet?

Do you feel comfortable and settled in your home, and never want to leave? Or are noisy neighbours, boisterous children and barking dogs forcing you to reconsider your postcode?

Are you able to enjoy playing your guitar or drums, disco dancing with the kids or singing your morning melody? Or are you always worrying your neighbours will be banging on the wall in protest?

Life is noisy, but environmental noise, neighbours and animals, and your own desire to make a racket, shouldn't stop you from enjoying life to the full in your own private oasis.

Over the last fifteen years, I've worked with hundreds of homeowners struggling with unwanted noise in their home – people who just want a quiet life or people who don't want to disturb their neighbours. I've found that most of these people are suffering from:

- **Poor health:** Noise makes them annoyed, stressed and irritable; affects their sleep, and their physical and mental health; and puts a strain on their relationships with anyone around them.

- **Lack of privacy:** Their home is not the private sanctuary it should be. They are constantly disturbed by hearing every aspect of their neighbours' daily lives; this in turn makes them very aware of what they do in their own house and what people might think of them, so they have no freedom in their own home.

- **Feeling unsettled:** Their house doesn't feel like home. They try different things – new home furnishings, going out more to avoid the noise – but no matter what they try they just can't seem to settle.

Many homeowners are struggling; they have no privacy, no place to retreat to from the outside world, they are constantly disturbed by their neighbours and this makes them wonder about moving. This unwanted noise affects their mental and physical health, leading to poor well-being and a lower quality of life.

If that feels like you, then I know how you're feeling. From the very first night my girlfriend and I got the keys to our first semi-detached house on a nice little *cul de sac* we realised we had a problem. We could hear the neighbours. We knew what TV programme they were watching, when they went to the kitchen to make a cup of tea, what time they went to bed and whether they brushed their teeth. We could also hear them talking and their dog barking. During the whole excitement of buying our first home together, I had never once thought about noise or noise from the neighbours.

I started to have trouble sleeping. I couldn't relax in our new home, and I was constantly disturbed by the neighbours. I began talking quietly when I was in the house: if I could hear them, they must be able to hear me. The saddest thing was that these were lovely people: they weren't having wild parties or shouting at each other until 3am. They were just a normal, busy family, living their lives.

This is where my obsession with soundproofing started. I decided to investigate and learn from the soundproofing experts. I was in the early stages of my building career. I thought I knew how to soundproof a house, and I had years of structural knowledge from building home extensions, barn conversions, installing different insulation systems, wall coatings, stud frames, suspended ceilings, drywall systems and floor systems. You name it, I had done it. My obsession with soundproofing research led me to the conclusion that the problem was the construction of the house. I knew that this was not unique to me, and that millions of people all across the UK must be suffering from unwanted noise in their homes. So, what could I do?

I managed to make our home noise free by perfecting the soundproofing in all the rooms. After I soundproofed my house and made it a home, it felt amazing. I looked forward to coming home; I felt settled and comfortable, like I had no reason to ever move. Even better, I could make as much noise as I wished without bothering the neighbours and indulge my passion for cheesy dance music.

There's no need for you to suffer. I turned the techniques I used for my own home into a unique four-step soundproofing method that works for every home. I'm going to share this with you right now.

Let my obsession be your liberation from noise pollution.

Imagine coming home after work and not noticing your neighbours. Watching the cars and trucks zip past and not hearing the noise from the traffic. Playing your music as loud as you want, whenever you want, in your own home without annoying others.

Soundproofing your home can give you all this and much more. As modern living gets louder, our lives are getting busier and construction techniques, budget and speed pressures shape the new housing stock; unwanted noise in the home is a growing problem. If you think you've tried everything, and there's nothing left to do but move house, wait! There is a solution.

A home becomes noise free when unwanted noise is no longer disturbing the homeowner; when the homeowner can make as much noise as they want confidently knowing they are not disturbing others. If you're a homeowner considering soundproofing as a way to reduce this unwanted noise in your home, then installing it correctly is vital.

The Noise Free Home will show you the four steps required to provide relief from unwanted noise. Following these steps will give you your privacy back. For many homeowners, something as simple as not hearing the neighbours' conversations and TV makes them feel comfortable in their own home again. Maybe it's having guests to stay without feeling embarrassed, or being able to work from home in peace and quiet, or no longer needing to sleep with ear plugs in.

A soundproofing investment will give you privacy in your home, feelings of well-being and you won't be driven to consider moving again. Our four-step method can effectively protect any domestic home from unwanted noise. My company, Quietco, is a domestic soundproofing company. We install sound insulation to walls, which reduces noise; design and build unique multi-purpose soundproofed rooms where the family can enjoy making lots of noise together without bothering their neighbours, which gives any home a massive wow factor; and have reduced noise for hundreds of people over the last decade.

We accept that soundproofing your home is a large investment, a decision that needs a lot of research and understanding to make. Read on to find out whether to move, or stay and soundproof your house.

The four-step soundproofing method

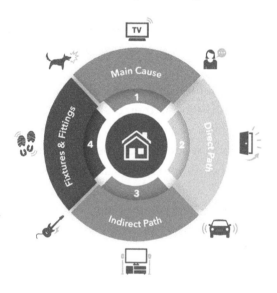

The four-step soundproofing method

To soundproof any domestic home:

Step 1: Treat the main cause of the problem: Fix the main cause of the noise getting into your home (see Chapter 4)

Step 2: Treat the direct noise paths: Apply soundproofing to the direct noise paths along which noise gets in (see Chapter 5)

Step 3: Treat the indirect noise paths: What indirect paths does noise follow: areas where it can come around the main structures? (see Chapter 6)

Step 4: Treat the fixtures and fittings: Make sure fixtures and fittings do not compromise the soundproofing installed (see Chapter 7)

'Treat' means fix, apply soundproof insulation or employ soundproofing techniques to address these problems. **All** these steps need to be completed if you want to reduce unwanted noise in your home **significantly**. Let me show you how.

PART ONE
YOU DESERVE A QUIET LIFE

1
Is Your House Built To Make The Noise Worse?

When you fly into the UK in daylight hours, have you ever noticed how close we all live together? Have you ever noticed how many attached houses, ie terraced and semi-detached houses there are? And how many houses are built next to busy roads and flyovers? Nearly 54% of the English housing stock is either terraced or semi-detached and 20.7% is apartments,[1] a great many of which do not have adequate sound insulation. Although sound insulation has now been included in British building regulations, I question whether these minimum standards really are fit for purpose.

---

1   See NHBC Foundation, '40 facts: Homes, housing and house building today' (2018), www.nhbcfoundation.org/wp-content/uploads/2018/09/NF81-WEB.pdf, accessed 13 October 2020

Ultimately the rapid development of noisy electronic gadgets and appliances, together with modern living, diverse employment opportunities and more working from home, means homes are noisier than ever.

I decided to completely understand noise in my own house and identify where the noise was coming from. I became obsessed with the construction of my house and knowing from what material every part of it was made. I identified each material's density, what materials are the present-day equivalent and their sound-insulation qualities, the size of the material, and how sound affects different materials and reacts as it interlocks with other materials. I became passionate about how different types of noise are transferred in houses and around buildings. I started applying this material understanding to every building project I worked on. Soon I could walk into a house, tap the walls and have a good idea what materials had been used and how the house was constructed and, from this, identify where noise was transmitting. Renovating my own property – with open-plan modern living, installing hard wooden floors and recessed downlighters, equipping it with the latest gadgets and working from home – gave me an understanding of the effects of neighbour noise and appreciate the discomfort it can cause to any homeowner, and this understanding has been instilled in Quietco's company culture and values.

Noise

Noise is sound that is created when an object vibrates fast enough that it produces a disturbance in a gas, solid or liquid; this generates a pressure wave in the surrounding air. Sound is the term used to describe the vibrations as they enter the ear and are then interpreted by the brain.

In this chapter, I am going to give a brief introduction to sound and theories of sound in relation to your home. This will give you a better understanding of unwanted noise and help you implement the ideas and concepts outlined in this book.

Humans' sensitivity of sound is affected by three things:

• The frequency (or 'pitch')

• The loudness of the sound

• The length/duration of a sound

Frequency

Frequency or 'pitch' is measured in hertz (hz). Sounds with frequency as low as 20hz can be felt by humans rather than heard. A teenager with good hearing can hear a frequency as high as 17,400hz. As we get older, we find it more difficult to hear higher frequencies; a normal, middle-aged adult hears up to between 12,000 and 14,000hz. For this reason, some convenience shops have started installing devices that emit high-frequency sound that is only heard by teenagers, to discourage teenagers from hanging around inside or outside the shop.[2] The voice of a normal male adult will range between 85 and 185hz and a normal female adult's voice between 165 and 255hz. The human ear is most sensitive at frequencies between 2,000 and 5,000hz.

Most sounds we hear from neighbours occur within 125 and 5,000hz. Most present-day sound systems can produce low-frequency noise below 80hz and some people have deep voices that will transmit through a house's structure. At present in the UK, government testing requirements for separating walls/floors to meet Part E of the Building Regulations only test in the frequency range 100hz to 3,150hz.

The good news is that 8,000hz sounds like a very high-pitched squeal, and most separating walls, ceilings or floors will barrier frequencies this high. These higher frequencies are actually quite easy to keep out through airtightness. High frequencies produce small sound waves that pass through small holes in the structure of your home. Low frequencies produce large sound waves that struggle to get through small

---

2   Kirk, T, 'The use of sonic "anti-loitering" devices is breaching teenagers' human rights', *Independent*, 2017, www.independent. co.uk/news/long_reads/use-of-sonic-anti-loitering-devices-breaching-teenagers-human-rights-a7880781.html, accessed 9 November 2020

gaps in the structure. Very low frequencies, between 20 and 100hz, travel the furthest and are the hardest to stop.

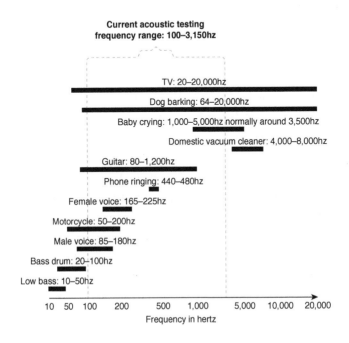

Frequency chart

Audible frequency range

The human range is commonly given as 20 to 20,000hz, although there is considerable variation between individuals.[3]

---

3   For example, see Abbott, R, 'What are noise contours?' (2018), www. bocaairport.com/faq-items/what-are-noise-contours, accessed 13 October 2020

When you turn up the bass on your subwoofer, you are filtering out the high frequencies and receiving more of the lower frequencies. When you turn up the treble, you are receiving more of the higher frequencies.

Airborne noise

Airborne noise is noise transmitted through the air, such as TV or radio programmes, music, conversation. Airborne noise is regulated in the UK by the government to tackle the noise pollution in newly built homes and refurbishment projects, through the Building Regulations Approved Document Part E. Builders and developers are required to install sound insulation to the walls, ceilings and floors separating some refurbishments and newly built homes and then have an acoustic test carried out. If the building passes the acoustic test, then the new development will be on its way to meeting building regulations and being signed off. Just like an electrical certificate, the acoustic certificate is part of the inspection process.

MEASURING SOUND LOUDNESS

Decibels: A decibel (dB) measures how loud a noise is.

50dB is quiet while 140dB is so loud it can immediately injure your ears.

The level of sound prevented from transmitting through separating walls and floors is measured

IS YOUR HOUSE BUILT TO MAKE THE NOISE WORSE?

(airborne sound insulation), and the higher the figure the better (the regulations require a score of at least 45dB DnT,w + Ctr – see Appendix for what this term means). The testing covers a wide range of frequencies, but unfortunately takes the average of the room. An amazing wall system on the party wall might achieve 70dB DnT,w + Ctr but if you have not treated the suspended floor void, the ceiling void or other holes in the structure, these would let neighbour noise through. If the sound test found the overall sound insulation average of the room to be 50dB DnT,w + Ctr, which is a pass, you might still be able to make out words and follow conversations from next door. This is explained further with the case study of Jane and Rob's semi-detached property in Chapter 8.

The acoustic test came into force on 1 July 2003 for refurbishment projects. To give you an idea what 45dB DnT,w + Ctr means in real life, in a 1950s three-bedroom semi-detached house, I paid an independent company to carry out a UKAS-accredited sound insulation test of the separating party wall *before* any soundproofing was installed. The sound insulation value of this wall measured at 47dB DnT,w + Ctr, 2dB *higher* than the building regulations require for new dwellings, so better sound insulation than the standard required for properties newly built today. However, I stood in our client's neighbours' dining room, the other side of the party wall, and managed to turn on our client's smart speaker.[4] After soundproofing using the

---

4   See www.quietco.uk/videos

four-step method, we achieved 62dB DnT,w + Ctr; this is a 15dB reduction in noise from next door. To give you an idea, 10db is a halving of sound as we hear it so the 15dB reduction represents a significant reduction.

Don't worry about what these technical terms mean in detail. That's what I'm here for.

Impact noise

Impact noise or structure-borne sound is most commonly associated with footsteps, for example footsteps on a wooden floor heard in the room below. Other examples of impact noise heard in the home are doors slamming, children running and jumping around, plug sockets and light switches, people using chopping boards, pots and pans clanging on a work surface, curtain and blinds being drawn, creaky floors, and bangs against a wall or floor or ceiling, ie where the noise transmits through the structure.

When an object hits the structure of a building, the sound energy is transferred to the structure and can then transmit through it, becoming a structure-borne sound. So, when your neighbours' door slams, you are hearing the noise of the impact of the door leaf hitting the door frame as the door is closed. You may also be hearing the door latch (the metal parts of the door) hitting the strike plate on the door frame. Both these impact noises transmit through the structure of the building all the way to your property. I outline a solution to door-slamming noise when we address the fixtures and fittings in Chapter 7.

In the early days, impact noise was measured by a steel ball being dropped from different heights onto a wooden floor; the higher the drop the more audible it would be in the room below. Later a 'tapping machine' was developed (see below). The strict testing practices, standards and methods acoustic engineers have to abide by are outlined in the British Standard for measuring impact noise, BS EN ISO 140-7. The tapping machine uses a series of steel hammers tapping ten times a second. This machine is placed on the sub-floor / temporary floor (not a finished or tiled floor) in the source room to be measured; acoustic engineers then measure the calibrated sound pressure in the room below. Impact noises can be reduced by isolating the surfaces from the structure, using floating floors, floating walls and suspended ceilings.[5] Approaches to treat impact noise are outlined in Chapter 5.

The key things you need to understand about the two different types of noise is that *both* impact *and* airborne noise *will* get *into* the structure and transmit *through* the structure of the house. You can make a room airtight but still hear music, TV, and conversation

---

5   The impact sound pressure level is called LnT,w. A maximum
    level of 64dB LnT,w for impact noise transmission to separating
    floors also came into force on 1 July 2003 for converted properties.
    Confusingly, the lower the LnT,w figure, the better. Since 1 July 2004,
    the regulations require airborne sound barriers to be a *minimum* of
    45dB DnT,w + Ctr for separating floors, ceiling and walls, and impact
    sound transmission to be a *maximum* of 62dB LnT,w.

noise from next door because it is transmitted through the structure. These noises will still get through unless you *isolate* the finished surfaces from the structure. There is no point in just applying sound insulation to a separating party wall if noise is getting in through the structure: you must apply soundproofing to any area that is connected to the source of that noise. More on this when we look at the main reasons soundproofing installations fail, in Chapter 6.

That's about as technical as we get in *The Noise Free Home*. Some key points are outlined above which build foundations for understanding how noise enters or leaves your home. Sound testing is worth understanding if you live in a new-build property or are thinking of moving to any house built in 2006 or more recently. It might be worth referring back to this chapter as we go through the different concepts and ideas outlined in later chapters.

What if this noise is caused by the construction?

The types of building materials used in the construction of a house have an impact on noise transmission. Some construction materials allow different types of sound to travel easily through them and some construction techniques inadvertently amplify the noise. Unfortunately, architects, builders and developers are unknowingly applying building materials and techniques that have been proved to make unwanted noise worse in homes.

For example, roofing techniques may do nothing to barrier environmental road noise or air traffic noise from getting into the house; in fact, commonly used materials and techniques can actually channel road noise around the house. The frustrating thing is that readily available alternative materials that are not difficult to install can reduce the noise from outside.

Developers and builders converting and renovating properties are unknowingly amplifying the noise from neighbours. They hide missing bricks or holes in the structure with a piece of plasterboard. They box in rolled steel joists (RSJs) the way they were taught, and they use lightweight, 'cheap' materials without understanding how sound transmits through them. 'One of the best materials for carrying sound waves includes metals such as aluminium… Sound travels at one of the fastest rates through aluminium, at 6,320 meters per second'[6] and aluminium stud frames are commonly used in modern developments.

A very widely used plaster-boarding technique known as 'dot and dab' has been proven by the National House Building Council (NHBC) to amplify structure-borne noise (see Chapter 4). My engineers

---

6   See Allan, S, 'Which materials carry sound waves best?' (2018), www. sciencing.com/materials-carry-sound-waves-8342053.html, accessed 13 October 2020

have carried out the same experiment and match the NHBC engineers' graph exactly.[7] This noise could come from neighbours, from a busy road or even be noise you make yourself. For example, you might keep your TV volume on low but still receive complaints from the neighbour; or your neighbour might be playing music at a low, comfortable volume but to you, sitting next door or downstairs, it sounds like they are having a party. The main cause of the noise could be your house. It could be a few different things but most likely the plasterboard technique or a hole in the structure from how it has been constructed.

UK house construction – new-build problems

In your brand-new house, you hear all aspects of your neighbours' life: conversations, TV, doors banging, toilet flushing. Despite this, your brand-new house meets current building regulations.

Other noise problems from newly built-homes that drive people mad are creaky floors, room-to-room transfer of noise and road traffic noise. One question many people ask me is, 'When the TV is on downstairs in the living room on the ground floor... why can I hear it on the second floor?' This is caused by a combination of different building techniques and the

---

7   You can find this report on www.quietco.uk/blog/nhbc-report-on-dot-and-dab-plasterboard

materials used. This will be explained in more detail in Chapters 4, 5 and 6.

What will I hear in my new-built home?

Just to be clear, there are some fantastic new home builders out there; however, if you google 'noise in new build homes' you will see which housebuilders have bad reputations. Unfortunately, these badly built new homes are giving new-build properties a bad name and the regulations too.

Although the threshold of 45dB DnT,w in the current Part E Building Regulations is better than having no regulations at all, if your new-build's rating is less than 50dB DnT,w you will most likely hear your neighbours' conversation and I would not be surprised if you could make out words. If you don't, then you have nice considerate neighbours who probably hear everything you do and are probably tiptoeing around their own house like mice. Modern living, with huge flat-screen TVs installed on walls, wireless speakers, surround-sound systems, subwoofers, together with flexible working hours etc, means that this test is not currently fit for purpose. Many of these devices produce frequencies that are outside the range of the current acoustic testing (100hz to 3,150hz). For example, most subwoofers on a surround-sound system have a recommended crossover frequency of 80hz. It is also worth noting that a male voice has a frequency range of 85 to 180hz, and a baby's cry is normally around 3,500hz but can range from 1,000 to 5,000hz.

Many new-build or refurbished houses and apartments that are full of plasterboard and lightweight thermally efficient aerated blocks are too expensive to soundproof. Some detached, semi-detached and terraced houses would need to be completely knocked down, or at least completely gutted, in order to soundproof them.

In Chapter 4, I explain how all the plasterboard in a new-build property needs to be removed, not just the board on the party wall. Construction that uses aerated inner blockwork and fills cavities with foam insulation means that noise transmits readily through the structure. Any soundproofing system to be installed therefore has to be isolated from the blockwork, taking up much-needed space in a living area that may already be small.

Research by housing charity Shelter and YouGov found 'half (51%) of new home owners said they have experienced major problems with their properties including issues with construction, unfinished fittings and faults with utilities.'[8]

---

8   Shelter, '"Rigged" housebuilding system means eight in ten families cannot afford new home, says Shelter' (2017), https://england. shelter.org.uk/media/press_release/rigged_housebuilding_system_ means_eight_in_ten_families_cannot_afford_new_home,_says_ shelter, accessed 24 November 2020

Loopholes

CASE STUDY

Dave and Debbie purchased their new home off plan and, like a lot of people, did not consider noise before moving in. Their semi-detached, three-storey property was built by a well-known housing company specialising in new-builds. The estate was still being built as they moved in and a few months later a family moved in next door.

Dave and Debbie got that horrible sinking feeling in their stomach when they discovered they could hear their neighbours in their dream house. Debbie could lie in bed and hear everything from next door; she could even hear whether the neighbour had washed his hands after going to the toilet. They could hear all the usual noises – TV, music – and could make out conversations word for word. They decided to make a formal complaint and nine months later an accredited sound test was carried out. The well-known independent organisation that carried out the sound test could only test in the kitchen, which backed on to the neighbours' lounge, because the kitchen was the only place where areas classed as 'a habitable room' could be found on both sides of the separating party wall: all Dave and Debbie's other rooms backed on to the neighbours' toilets, bathrooms, storage cupboards, stairs etc.

Dave and Debbie were fuming after waiting so long for this sound test. They did not hear the neighbours in the kitchen. They heard them in their living room, in the bedrooms, in their bathroom, on the stairs and in their downstairs toilet.

The sound insulation test was passed with flying colours, of course, and the case was closed. Dave and Debbie were left with hearing every aspect of their neighbours' daily lives. Debbie was a light sleeper and was constantly woken up by the neighbours getting up in the night to use the toilet, and their irregular shift patterns. The noise and stress went on for two years before the couple moved, leaving the problem to some other poor young couple excited about their new home.

Housebuilders quickly realised that soundproofing is quite expensive and installing sound insulation in a lot of new-build homes would reduce the profit margin and take up space. Also, who would install it? These builders have not been taught these techniques in their apprenticeships; on their drawings, architects quote vague soundproofing statements taken from an acoustic supplier's website, with a caveat to protect themselves. There must be another way.

The practice of backing habitable rooms on to inhabitable rooms in houses, so that habitable rooms do not adjoin, is a very old construction technique. Done correctly, it can be good practice, to try to reduce noise disturbance between main living spaces. For example, on one side of a separating party wall you might build living rooms and on the other side stairs with under-stairs cupboards, toilets, cloakrooms or small hallways. Upstairs you might build bedrooms and on the other side of the separating party wall *en suite* bathrooms. Similar floor plan layouts are also used for new apartments and flats.

When you apply this construction technique to modern home building, it creates an opportunity to meet the acoustic testing regulations with minimal sound insulation and effort. With any accredited testing, such a layout makes it difficult to get accurate readings and, even worse, the acoustic testing equipment cannot be set up on stairs, in a bathroom or toilet or in cloakrooms.

In many cases like that of Dave and Debbie, this would only leave one room in the entire house that could be tested.

Developers purchasing properties and converting them into flats and houses seek a return on their investment, so it is important to keep project costs as low as possible. For most developers, the main focus is a cheap build and nice-looking fixtures and fittings. They know what your eye is drawn to – nice tiles, hardwood floors, downlighters, open-plan living – they know what sells a house. Any additional sound insulation that is better than the minimum required seems to be considered a waste of money.

The government sets targets for the number of homes built, so housebuilders are under pressure to build as quickly as possible. This has no doubt encouraged a culture of 'who can build the quickest' and reduced the quality of home building in the UK.

Innovative building methods, such as prefabricated roofs and walls, have been widely tested and are used more and more in new-builds. Some companies now claim to slash months off the time required to build a new home by manufacturing houses in a factory.[9] These prefabricated houses will be delivered to development sites already fitted with kitchen and bathrooms. This is great for the housing shortage and people that need a roof over their head. However, it will be interesting to see how these prefabricated houses perform acoustically, since they often lack mass and sound insulation compared to traditional methods such as concrete blocks.

Summary

I appreciate the stress and anxiety that unwanted noise can cause any person affected by it. By soundproofing my own house, I was able to experience the total satisfaction of overcoming that noise.

Building regulations for acoustic testing only cover frequencies between 100 and 3,150hz and do not test all the different frequencies that neighbours (and your own family) can generate in the home. The two different types of noise will get *into* the structure of the house and transmit *through* it. You can make a room

---

9   Collinson, P, '"UK housebuilding revolution": £65,000 prefab homes go into production', *The Guardian*, 2018, www.theguardian.com/society/2018/nov/30/uk-housebuilding-revolution-65000-prefab-homes-go-into-production, accessed November 24 2020

airtight but still hear music, TV or conversations from next door unless you isolate the finished surfaces from the structure-borne noise.

Low-frequency sound, below 100hz, whether from a busy road or a neighbour's voice, can transmit both through the air and structurally. To barrier low frequencies, you must isolate your finished walls, ceilings and floors from both airborne and structure-borne vibration.

We have proved that you can still hear conversation and everyday noises through party walls even in houses that meet current building regulations. Even the Local Authority Building Control (LABC) website identifies noise as a problem in new-builds.[10]

The construction of a house (building materials and the techniques used) can make noise worse and can be the cause of your noise problem.

The old practice of building attached houses so habitable rooms do not adjoin is good practice, to try to reduce noise disturbance between main living space, but developers appear to be applying this technique to reduce the amount of testing between separating party walls. The culture of building houses quickly and cheaply has to change in the construction industry.

---

10 For example, see www.labcwarranty.co.uk/blog/reduce-noise-in-your-new-build-with-the-help-of-british-gypsum, accessed 13 October 2020

The quality of new-build houses must improve, through the use of higher-quality materials, better workmanship and responsible, sustainable building techniques.

You could not have known your house construction, or your new extension or complete home renovation, would be the main cause of unwanted noise. Building regulations do not go far enough. It is not just your house; this problem affects millions of homes across the UK. There are ways to build better.

2
Is Your House Making You Ill?

When a homeowner struggling with noise approaches us, they often feel annoyed and stressed from the noise; it affects their sleep and their health and puts a strain on their relationships with anyone around them. They have no privacy in their home because they are disturbed by the noises of their neighbours, which also makes them conscious of what they are doing in their own house and what people may think of them.

When faced with any noise problem, many people find themselves wondering:

- Should I move or should I soundproof my home?

- What result will I get if I stay and soundproof my house?

- Will this actually work?

CASE STUDY

Lynne and Dave loved the area and their street; the house was a 1930s semi-detached with large rooms and had everything they needed. The only problem was the noise from the neighbours, who had put a TV on the chimney stack on the other side of the separating party wall. The TV was heard in every room along the separating party wall, in particular the rear dining room and the back bedroom above. As well as TV noise, Lynne and Dave would hear conversations and a dog that howled whenever it was left alone. In the rear dining room, in particular, they could hear the sound of the dog's feet on the neighbours' hard wooden floor. Because they could hear so much of their neighbours' daily lives, Lynne and Dave felt that they had no privacy and were embarrassed to have friends round; and Lynne was worried about having the grandchildren visit in case they made too much noise.

Due to the noise, Lynne and Dave had started looking to move to a new house, 'something they did not want to do at their age', and this made them feel really unsettled in what they had thought would be their final home. They worked out the costs of moving but could not find anywhere they would be happy to call home. They figured their house had everything they needed; 'It's just the noise!' Lynne said. After discussing the four-step soundproofing method with us, they addressed as many key areas as they could afford. To limit disruption, the upstairs room was treated in the first week, giving Lynne and Dave a quick win. After installing soundproofing downstairs, including a new carpet, they soon realised that they no longer heard the noises that made them uncomfortable in the two back rooms.

Lynne and Dave are pleased they soundproofed their home. It has been a massive relief of stress for them; Lynne feels like a weight has been lifted from her shoulders, as they can now stay in their 'forever home'. Doing one room first allowed them to gauge the reduction in noise they would get from soundproofing the whole house. Not being able to hear the neighbours' dog howling or their TV was very reassuring and they look forward to soundproofing the front of the house in due course.

___

Let's go back and start at the beginning, with a detailed look at what noise in the home *is*. Noise is unwanted sound - 'sound' is the physical component and 'unwanted' describes the emotional and psychological effect that noise has on a person. Home means different things to different people, but it is often described as the most personal and sacred place in which to shelter from the rest of the world. This is why many people find noise entering their home so intrusive and irritating. Not being able to relax in your own home, not being able to read a book or watch your favourite TV programme without hearing your neighbours or worrying that they might be hearing you, is just depressing.

When you constantly hear unwanted noise at home, it can cause stress and anxiety, raise your blood pressure, change your mood, and even put you off your dinner. Noise affects different people in different ways, but I also believe that different *types* of noise affect people in different ways.

How is the noise affecting you?

The considerate neighbour

You may be that considerate neighbour who is very conscious of the noise you make, whether you are simply aware of your own general living noise, or hear your neighbour, or are conscious whenever you play your music or TV, or have people round. Would you love to be able to play your music as loud as you want whenever you want? Do you play an instrument or have a noisy hobby that you want to fully enjoy in the privacy of your own home? Do you feel you can't, for fear of bothering the neighbours? I have met people that have lived this way for many years, and it does affect their mental health and relationships.

Environmental noise

Noise entering your home from the street through your windows and roof – for example a neighbour revving his car unnecessarily, a road that is busy during rush hour or has a constant 'whoosh' of cars driving past at all hours of the day, the occasional motorcycle or lorry driving past – produces a constant reminder that there's a road out there. Other homeowners may suffer from the hum of a motorway, dual carriageway or railway track.

Neighbour noise

Some homeowners find environmental noise less annoying than voices, such as conversation and laughter coming through the party wall or ceiling, for example. A neighbour playing heavy bass music at 11 o' clock on a school night might be even worse. Hearing impact noise – neighbours' stairs, footsteps through a ceiling – constantly throughout the day, whether in a communal hallway or through a party wall, does not bother some people. But the tone of voice of a particular person who lives above them, or a cough, or a neighbour clearing their throat, can infuriate some people.

Some neighbours enjoy listening to the neighbours' children laugh and play together, whereas other neighbours cannot tolerate the noise of children and spend thousands of pounds blocking that particular noise out. Before you invest in sound insulation, try to think what different sounds affect you.

For me, it was loud music from the neighbours, almost making out the songs. Before soundproofing our house, their music could be heard in every room.

If the neighbours decided to have a little dance about with low-frequency dance music, there really was no escape from the noise, whatever I was doing. It could happen at several times of day. Working from home became very difficult: I could not concentrate or take professional phone calls. The neighbours are a normal, busy family and should be able to play music when they feel like it, at reasonable times. The construction of our house was the problem; the only sound insulation it had was the brick wall itself, which was insufficient.

What effect does noise from neighbours have on you and your health?

Environmental noise can be consistent, or at least predictable, and therefore can easily be measured by acoustic engineers, who set up devices over twenty-four-hour periods and then analyse the data for particular types of noise. The data can be used to inform decisions on planning a new development, new roads or railways, or wind farms, for example. Noisy neighbours, on the other hand, are rather more unpredictable and very difficult to measure. Consequently, residential noise (particularly the effect of noisy neighbours) has not received a great deal of research. Diana Weinhold (2015) outlines in her report 'Sick of noise: The health effects of loud neighbours and urban din' that there is no doubt that noise affects health. For example, emotional and psychological effects increasing stress, leading to headaches, fatigue, disrupting sleep or, worse, joint and bone problems, lung disease, cardiovascular problems and heart attacks.

Hypertension is a common effect that noise from neighbours has on a person's health. I have experienced this personally and I have also met hundreds of people affected by this. Hypertension is abnormally high blood pressure and a state of great psychological stress. I have found that neighbour noise does not necessarily have to be loud or excessive to be an annoyance. Low-level noise can cause you to feel stressed and uncomfortable but still be at a volume (decibel level) that falls well within current building regulations. A slight irritant, such as the muffled noise of a neighbour's TV, conversation or sudden furniture moving, is all it takes to raise your blood pressure.

It is human nature to have your attention drawn to sounds like footsteps and the comings and goings of modern living environments. These types of sounds are information and can cause a negative emotional reaction. This dates back to the days of cavemen and women: early humans were instinctively aware of information noises and listened out for danger. Although nowadays few of us are listening out for tigers, this innate warning system still functions in all of us.

Often, clients who know that an irritant noise comes from neighbours get more irritated than those who do not know what the noise is or where it is coming from. Is this true for you?

Clients who confess to being 'sensitive to noise' normally have a few particular noises that they want to keep out: a wardrobe door, a washing machine, kitchen noise, for example. The client will have a heightened sense of hearing this particular noise because they know its source and can pinpoint what is happening, when and why. This knowledge tends to aggravate them and cause more stress and anxiety than an unknown sound might.

Space invader

Noise invades your space. Even if you have a large house and your nearest neighbour is a mile away you will still be invaded by noise: perhaps someone chain-sawing logs in the distance, a tractor ploughing or spraying a field, or a distant road or the low-frequency hum of a motorway. All these noises can travel for miles within the right landscape.

If your neighbour suddenly starts playing loud music, which you can hear in every part of your own home, your personal space has been invaded by your neighbour. Your home, your personal and sacred shelter, is no longer a shelter, no longer a place to relax or escape. What do you do?

English law defines a nuisance as 'an unlawful inter-ference with a person's use or enjoyment of land or of some right over, or in connection, with it'. For noise to amount to a statutory nuisance, it must be 'prejudicial to health or a nuisance'.[11]

Noise can invade your personal space at any time, can spoil any event, and affect your mood and those of others around you. More than just annoyance, anger or frustration, the negative emotional reaction can be feelings of helplessness, withdrawal and even depres-sion. Noise can cause you to change your plans and this could happen at any time of the day. Imagine this happening to you weekly, or even daily; this would no doubt affect your mental health.

CASE STUDY

I met a lady that suffered so badly that she had a nervous breakdown. The stress and anxiety led to a number of different health problems and she found herself in and out of hospital on a regular basis.

Annie and Bill live in a beautiful, four-storey semi-detached house in the centre of Leamington Spa. Next door was a house of multiple occupancy (HMO). Before Annie and Bill moved into their newly renovated house, the builders that carried out the renovation promised it had been soundproofed. Unfortunately, the noise from the tenants next door was all the usual neighbour noise

---

11   Section 79(1)(g) and (ga) of the Environmental Protection Act 1990, www.environmentlaw.org.uk, accessed 17 December 2020

– TV, conversation, music – but also banging doors and coat hangers in the wardrobes on the other side of Annie and Bill's main bedroom wall.

We quickly identified that the builders had thought that, by using blue, sound-resistant plasterboards and acoustic mineral wool, they were installing soundproofing. Unfortunately, the techniques and the combination of other materials they used channelled the noise from the tenants into every room in the house. In essence, they had made the noise worse. To solve the problem completely would require removing the plasterboard and treating every wall, ceiling and floor in every room in the house, which was unaffordable. This information was difficult for Annie to hear and the realisation that the builders had made things worse caused more health problems.

The builders had also removed a chimney stack and left bricks missing in the party wall, leaving holes going through to next door.

We gave Annie and Bill complete peace and quiet in their bedroom. Being able to go to bed when they wanted and wake up when they wanted made a huge difference to their well-being. However, years later, Annie has still not fully recovered from the ordeal.

As anyone in the soundproofing industry will probably tell you, it does feel quite amazing to be able to make such a difference to people's lives; to improve their life and well-being is a very rewarding part of the job.

I think people are not aware of the full costs of intrusive noise in their home and the effect it has on their health. They may consider the noise a nuisance but do not fully understand the possible long-term consequences for their health. Intrusive noise, be it road noise or noise from neighbours, causes sleep disturbance and stress, leading to health problems. The biological mechanisms that release the stress hormone cortisol in your body can be triggered by the type of noise, its volume, frequency and timing, all of which will affect your body in different ways.[12] Different people are affected by different noises. In most cases, noise from neighbours will result in some kind of stress, leading to lack of sleep. As soon as the noise starts affecting your sleep, seek help.

Cirrus Research has been a world leader in noise measurement instruments since 1970. Over the years, it has collected valuable data relating to domestic noise, finding (for example) that:

- 1 in 10 Britons (approx. 6.5 million people)[13] claim they are regularly disturbed by noise from neighbouring properties

- 28% have complained to their local council

---

12   Weinhold, D, 'Sick of noise: The health effects of loud neighbours and urban din', Grantham Research Institute on Climate Change and the Environment, Working Paper No. 213, October 2015, www.lse.ac.uk/GranthamInstitute/wp-content/uploads/2015/10/Working-Paper-213-Weinhold.pdf, accessed 12 October 2020

13   See Cirrus Research, '6 steps to resolving noise nuisance complaints', www.cirrusresearch.co.uk/library/documents/ebooks/6-steps-resolving-noise-nuisance-complaints.pdf, accessed 14 October 2020

- Less than 33% have complained to their neighbours directly for fear of confrontation

- 5% of those who did confront their neighbour were threatened and 1% had a physical fight

- The most common complaints come from loud TVs, barking dogs, parties, DIY and arguing neighbours

- People living in flats are the most susceptible to noise nuisance, with 10% claiming they had been left continually disturbed or stressed

Traffic and airport noise

I've shown that noise pollution can affect one's health, and that broken sleep can raise blood pressure. 'A recent review suggested that risk for cardio-vascular outcomes such as high blood pressure (hypertension), heart attack and stroke increases by 7 to 17% for just a 10dB increase in aircraft or road traffic noise exposure.'[14] Do you feel you can live with it? Is sleeping with ear plugs to shut out road noise not too bad?

---

14 Clark, Dr C 'Aircraft noise effects on health', Queen Mary University of London (2015), https://assets.publishing.service. gov.uk/government/uploads/system/uploads/attachment_data/ file/446311/noise-aircraft-noise-effects-on-health.pdf, accessed 17 December 2020

CASE STUDY

Steve and Natalie purchased a beautiful barn conversion in the Cotswolds, in a lovely village within walking distance of the local school and village shop. They did not have much of a front garden – the house was next to the village road, meaning noise from cars and tractors. Standing in the bedrooms, you could hear and feel the vibrations of tractors driving past, and without looking out of the small windows you could picture the tractor as it passed the house. You could make out words of conversation from people walking past, particularly during the school run. If a car drove past late at night, it would wake Steve.

They tried to live with the road noise for a month, before it caused so much stress and sleep disruption that they considered moving again. Steve and Natalie did not understand how the noise could come through the massively thick walls of the original barn. They started to research soundproofing and learnt that the materials and construction used on the roof during the conversion were amplifying the noise.

Soundproofing meant a huge compromise for Steve and Natalie, since sound insulation meant covering up some of the stunning oak rafters. I designed an approach that would reduce the noise and, although many rafters would have to be covered, we were able to leave exposed the big oak purlins from the original barn roof. Our ceiling system gave an amazing result for Steve and Natalie, who can now barely make out cars and farm traffic going past. When testing in the empty, newly plastered main bedroom, Natalie had to check through the window to see whether a car was coming. We made their main bedroom noise free, with such success that we were called back three months later to install the same system in the guest bedrooms. Steve and Natalie no longer need to move. They fell back in love with their barn conversion in the Cotswolds.

Putting it off

Arguably, stress and hypertension have health effects as bad as damp, mould and other building problems; yet, through lack of awareness of these effects, people put soundproofing on hold or put less importance on it. They put higher importance on a new kitchen or a bathroom renovation than on soundproofing.

If you own a property attached to another house, and have noticed damp, dry rot or a water leak that affected both houses, the neighbours, however unapproachable and awkward, generally would not hesitate to let engineers test the problem on their side if requested. What do you think your neighbours would say if you approach them and say, 'We have a noise problem on the party wall'? What many people say during neighbourly conversations is, 'Can I ask you a question... Do you hear us? Do you hear our TV?' The neighbours usually respond, 'No, not really, it's fine... we hear sometimes, but that's because we are attached... the walls are so thin.'

Maybe your neighbours are not that approachable. What do you think they would say? Is it a conversation you have thought about?

Noise transfer is a socially awkward subject. A lot more education and awareness will be required before it can be made compulsory to include a 'decibel rating' in the house prospectus. 'Neighbour noise annoyance is strongly associated with eight different physical and mental health symptoms'... 'pain or discomfort in the shoulder or neck, and in the arms, hands, legs, knees, hips or joints'.[15]

Summary

There is not enough research into neighbour noise, but there is no doubt that unwanted noise in your home can affect your health and put a strain on your relationship with anyone around you. It can make your house uncomfortable and not feel like home, and you may find yourself thinking about moving. Don't worry if you feel this way, it is not unusual to feel like this and there are solutions. If noise from neighbours is the problem, then, at some point, hearing every aspect of your neighbours' daily lives will make you

---

15 See Jensen, H.A.R., Rasmussen, B and Ekholm, O, 'Neighbour noise annoyance is associated with various mental and physical health symptoms: results from a nationwide study among individuals living in multi-storey housing.' BMC Public Health 19, 1508 (2019), https://bmcpublichealth.biomedcentral.com/articles/10.1186/s12889-019-7893-8, accessed 12 October 2020

aware of what you're doing in your own home and what people might think of you. All this affects your mental health and is made worse by lack of sleep. As soon as it starts affecting your sleep, seek help immediately.

Although different people find different noises annoying, it only takes a slight irritation, a small amount of unwanted noise to trigger hypertension, which stiffens your blood vessels. Whether it be noisy people living next door, a busy main road or not being able to make noise yourself… it does not have to be like this, there is another way.

3

Have You Tried Everything And Still Live With Unwanted Noise?

To 'soundproof' is 'to make a room or building resist the passage of sound and not allow sound to go through' a wall, floor or ceiling, for example. 'Soundproofing' refers to the construction of materials that prevent sound from passing, is often referred to more technically as 'sound insulating' and is different from sound absorption. Soundproofing is the reduction of sound energy as sound transmits through a barrier. Sound absorption is positioning material like high-performance foam to absorb sound energy, which is lost to heat. This book focuses on soundproofing your home using sound insulation to reduce unwanted noise.

Why do people soundproof their home?

People have a wide range of needs and wants, experiences, and expectations in relation to soundproofing. We get enquiries ranging from noisy neighbours through to soundproof steam rooms... in case you are wondering, some clients like complete silence in their steam room, to help relaxation or meditation, and others want to play relaxing music in their steam rooms without bothering anyone else.

Advances in technology bring noisier devices which, together with open-plan living and diverse employment opportunities, means noise in homes will get louder. The building techniques and materials currently used are just not suitable for today's new noise demands and simply need upgrading.

Other reasons to install sound insulation would be to compensate for poor workmanship, such as the use of plasterboard adhesive, or alterations that have weakened the original sound insulation, such as removing chimney stacks, installing uninsulated RSJs on the party wall for open-plan living or a loft conversion using modern materials; the list is endless.

A new neighbour may have a different lifestyle to the previous occupant. You may have just moved into your new home, next door to a quiet elderly neighbour who doesn't make a sound, apart from the occasional loud TV once a week. A year later the house is purchased by a family and completely renovated with open-plan rooms, carpet replaced with laminate floor, single light pendants replaced with recessed downlighters, and TVs on the wall... and then they get a noisy pet.

Many homeowners move into a new-build property and find that, although their new home meets current building regulations, TV, music and general living noise still come through. For example, some owners of new-build apartments contact us about slamming doors and similar noises that are not controlled by current building regulations.

CASE STUDY

Caroline and Duncan got their builders to install some 12.5mm blue, sound-resistant plasterboard to their new kitchen extension before they plastered it, and then installed a beautiful open-plan kitchen with granite worktops and LED lighting. Unfortunately, the neighbours complained all the time about the noise in the new extension. Caroline and Duncan found themselves talking quietly in there and always telling the kids to be quiet. The techniques and methods the builders used had made the noise worse, for Caroline and Duncan and their neighbours, too. They now realise that fixing the problem would require taking the entire kitchen extension back to blockwork, removing all the plasterboard, uninstalling the new kitchen and taking all the plumbing and electrical back to 'first fix' – see Chapter 7.

You may want to soundproof because, 'I didn't use to hear my neighbours – why do I suddenly hear them now?'

- Did new neighbours move in?

- Did you or your neighbours undertake a renovation?

- Have carpets been replaced with hard flooring?

- Has the structure been altered: chimney stack removal, extension?

- Have the fixtures and fittings changed: recessed lighting?

- Are rooms now used for different purposes?

- Have new devices been mounted on the structure: TV or speaker on the wall?

- Has an appliance been moved: tumble dryer, washing machine or stereo?

- Have lifestyle, job, routine, shift patterns changed?

All these are possible reasons for suddenly hearing your neighbours.

Some people believe that if they do not hear their neighbours, their neighbours cannot hear them. If told the property is up to building regulations and fit for modern living, they think that the neighbours are not bothered by the noise they make. They therefore continue making noise in their home. As a result, the neighbours are disturbed and talk quietly, the whole street learns their business, they are unaware they have no privacy and the house gets noise complaints. When they realise that the construction of their house is doing nothing for noise and they have no privacy, they are embarrassed and learn that investing in proper soundproofing will solve their problem. The results are amazing.

CASE STUDY

Colin and Rachel moved into a semi-detached house next door to an elderly couple. Seven years went by. Colin did not hear his neighbours so he assumed they did not hear him. The elderly couple were lovely and the relationship between the two families was very pleasant. Colin and Rachel started a family and it wasn't until the kids were running around making noise that the elderly couple mentioned to Colin that they could not only hear the kids, but hear everything Colin and his family did; they always had and accepted the situation.

Colin was taken aback to realise the neighbours had heard everything through the party wall all this time. Rachel and the neighbours tested this: she went next door and had a conversation. Colin was shocked at how much he could hear through the party wall.

He immediately wanted an approach to give him and his family privacy, and the lovely couple a respite from the kids. After a detailed consultation, Colin realised that it was better to treat the noise at source and agreed to have the soundproofing on his side of the party wall, with the elderly couple contributing to the cost. After the soundproofing was installed, the families decided to test it. No one could hear Colin shouting at the top of his voice from his house to the neighbour next door. Colin had a sweat on trying to be heard; everyone had a go at shouting as loud as they could; and the people next door heard none of it. The wonderful relationship Colin and his family had with the neighbours had been restored, along with his privacy.

If you have a lot of different neighbours over the years, you may be unaware of the noise you are making. Maybe you continue to play your music loud, have parties, shout and bang around: unwittingly you become the inconsiderate neighbour, not knowing you are making the neighbours' life hell. Relationships with neighbours get worse; noise complaints are made to the council and kept on record, affecting any future sale of both properties.

The funny thing is, if you asked your neighbour, 'Do you hear us?' what do you think they would say if they actually could hear you and all your conversations?

Common mistakes made by people with noise issues

Here are a few of the mistakes people with noise issues often make when trying to cure them. Having removed so much of other people's work in other people's homes, I thought it would be beneficial to share the most common mistakes in thinking and actions, hopefully saving you some time, money and heartache.

Many believe a general builder knows how to sound-proof. I strongly believe that no one is a jack of all trades; building and construction are so complex, with thousands of products, materials, methods and techniques. You wouldn't ask your dentist to treat your broken leg, your roofer to do your bathroom tiling or your painter and decorator to install your boiler. The same applies to soundproofing: it is a specialist trade, solving a complicated problem.

CASE STUDY

Julie was suffering from noisy neighbours. Shouting and aggressive arguments would transfer through the party wall of her end-of-terrace property. After retiring, she became quite ill and spent more time at home, and the noise from the neighbours became unbearable.

She thought a general builder would know how to deal with noise, so she had a builder apply soundproof plaster to the walls and then fix sound-resistant plasterboard to the entire party wall. All the rooms in the house were completely plastered, painted and decorated, ready for Julie to move back in.

Unfortunately, Julie could now make out words and follow conversations and arguments; her health deteriorated further.

The builder had made the noise worse. He had used the 'dot and dab' technique – applied large dollops of adhesive to the wall, positioned the sheets of plasterboard on those dollops and tapped them level to meet the existing surface. He *had* used sound-resistant plasterboard, thinking this would make the difference. Unfortunately, not only was the 'dot and dab' done incorrectly (the adhesive did not always cover 20% of the board), but the dollops of adhesive created pockets of air in which the sound resonated and was then channelled behind the plasterboard across the whole of the property, to parts of her home that had not been affected before.

She worried that removing what the builder had done, and having it done again by a soundproofing specialist, wouldn't be worth it. What if the result wasn't much better? After detailed consultation and advice, she realised that the approach the builder had taken had amplified the noise and how removing it and using proper soundproofing systems and installation techniques would solve her noise problem. With only 2 inches of space available, using four of the most advanced products on the market, the direct noise path was fixed (see Chapter 5) and the indirect noise paths were treated (see Chapter 6). After installation, the noise was almost eradicated, giving Julie peace and quiet in her house again.

Reading this far you may be thinking, 'Nope, I'm not convinced, Jim… seems like way too much disruption already. I'm not going to soundproof this house… I'm going to move!'

Will moving house solve the problem?

Although we receive a lot of calls and enquiries, and give out a lot of estimates to help people stay in the homes they love, not everyone decides to go through with soundproofing.

CASE STUDY

Blaming noisy neighbours and not the insulation of his house, John moved to a new house in the same area, closer to the local schools in a nice village near Oxford. Three weeks after he moved, we received a call: John had even worse neighbours. He had checked the property at different times of the day before purchasing, but nothing could have prepared him for the noise in his new house from next door. It was keeping his children awake; foul language and bedroom noises were very audible through the walls and the children were beginning to ask difficult questions.

Shocked by the loudness and clarity of what he heard through the party wall, John wondered how much of the extreme noise from the neighbours could be reduced by soundproofing. The noise was the only problem he had with his house. In the soundproofing consultation, we showed him that a builder had attempted to soundproof the property before selling it, and made the noise worse by channelling it into the children's bedroom, the kitchen and chimney stacks.

John decided to invest in soundproofing and is delighted. He has a happy family home in a great area and the children can sleep without being disturbed by the neighbours.

Remember:

- More often than not the construction and building techniques used on a property cause noise to invade your privacy.

- The neighbours don't have to be noisy to make you feel uncomfortable in your own home; general everyday noise can be intrusive.

- The same problem might crop up in the new home.

- When looking to move, look at the structural problems I have pointed out: plasterboard walls, recessed downlighters, an open-plan kitchen and uninsulated RSJs all transfer noise from next door.[16]

- Get advice on soundproofing before buying and then take any estimated costs into account in any offer you make.

- If you are looking to move just to escape noise, look at your own property construction first and see whether soundproofing might be a cheaper alternative.

---

16 If you would like more information on what to look for when considering moving, check out the '7 things to look out for' blog post at www.quietco.uk/blog/will-you-hear-your-neighbours-7-things-to-look-for

We deal with at least two cases every month in which people move away to escape the noise in their home. Unless you move to an old detached property, you will still hear your neighbours. If you move to a semi-detached or terraced property, you may have to soundproof at some stage. You may have a quiet elderly neighbour living next door now, but circumstances can change quickly. Moreover, just because you don't hear your neighbours it does not mean they don't hear you: they may hear every part of your daily life, and be very aware of what they are doing and the noise they make themselves. I have met hundreds of people that, for years, have tip-toed around their house not just because they are worried about their own privacy but because they are really nice, considerate neighbours.

I have already decided I will just move

If you decide to sell your house to escape 'noisy' neighbours, thinking it would be too much work to soundproof (even after reading this book), your solicitor will ask you to fill out a Seller's Property Information Form. On this form you will be asked to declare any disputes with your neighbours. At the time of writing this book, this declaration is still a grey area: what constitutes a noise dispute is open to interpretation. There have been many legal cases where new homeowners have sued the previous owners years after moving in, for not declaring that the neighbours are noisy when selling their home. Any noise dispute, which normally involves making a report to the council or local authority, will be picked up in the searches when selling your house. So make sure you declare it, because buyers could take action against you years after you sell your house.

In my whole career helping people with their noise problems, I do not recall one case where the problem was resolved by the local council. In my profession, I understandably do not hear about the good cases; but I know of many cases where councils and council officials have made the relationship between neighbours worse and escalated the situation. I would say that getting the local authority involved is something you need to do, to begin legal action against your neighbours for excessive noise or as a stepping stone. However, my opinion is that this could devalue your home. Discovery of a noise problem when buying a new home is a very good reason for buyers to pull out. It is therefore worth exploring soundproofing first.

Do-it-yourself soundproofing

A common mistake made by engineers or very capable do-it-yourselfers is to try to save money by doing it yourself. This is a very common mistake, but sometimes the only option for many families. With a passion for helping people with their noise problems, I always try to give as much advice as I can to those that want to try to DIY, mainly because my team and I hate taking it all down. Hopefully the ideas and concepts in this book will help those of you considering a DIY approach; you may be able to apply the ideas to improve the soundproofing result for you and your family. I am sure there are loads of successful soundproofing projects completed by very capable do-it-yourselfers, so please forgive me: I only hear about the bad ones. Maybe a textbook or training school should be my next project.

Some people think that doing home renovation themselves, including trying to soundproof, will save them money, so they buy materials online and give it a go. If they achieve a poor result, this often makes the noise worse, giving them no privacy and leaving them feeling more stressed, having wasted time and money and used up space in their home.

CASE STUDY

Danny is a perfectionist and very capable do-it-yourselfer. He and Jenny live in a three-bed semi-detached house in Warwick, a great location in a beautiful part of the country. Danny in particular became aware of neighbour noise and this made him very sensitive to the noise he was making in his own house. He found himself talking quietly and constantly telling the kids to be quiet, too. He would have the news on really quietly early in the morning and was never able to enjoy a good film because he was worried about the noise. Danny had done up a few properties in his time and thought he could save money by installing soundproofing himself. He applied a popular, heavily marketed product he purchased online to the entire lounge party wall, thinking this would solve the problem.

Danny realised he had taken a lot of space off his lounge, reducing the living space and not doing anything to reduce the noise, so he decided to have soundproofing professionally installed.

We took care of all the plastering, electrical, carpentry, finished flooring etc. Unfortunately, Danny had used an industrial-strength acoustic glue, so his soundproofing attempt had to remain behind the new wall system.

Danny and Jenny were very pleased with the new soundproofing, and we managed to test the result after installation. We put on *Transformers* the movie at a high volume through his new surround-sound system, then went next door and stood in the neighbours' lounge to see whether we could hear anything. To the family's amazement, no noise was heard next door.

Other bits of the house first

So you understand why soundproofing is important, and you are going to do it… but not yet – maybe after the bathroom or kitchen has been installed.

One very common mistake is to concentrate on other bits of house renovation first, thinking, 'We can always add soundproofing later, if needed'. Are these investments more important than soundproofing? Furthermore, does soundproofing add any value to your house?

CASE STUDY

Having been in the trade, Andy realised soundproofing to the party wall would make life easier in his house and give him more privacy from the neighbours. He had a clear plan of action: first, he wanted to install a new kitchen and bathroom, then renovate all the rooms.

Andy had his own team of builders sand all his floor boards and apply three layers of sound-resistant board to each wall, alcove and chimney stack along the party wall. Unfortunately, when he moved back in he could clearly hear the neighbours, louder than before.

After a detailed consultation, he realised that not only had all his 'soundproofing' got to be removed from the newly painted walls, but the finished sanded floors had to come up too. He rejected the four-step method at first, and requested only the separating party wall (the direct noise path) be soundproofed, without touching the sanded floors or perpendicular flanking walls. The soundproofing to the direct noise path, the party wall, made a small difference when he stood immediately in front of it, as expected, but this was not enough. Andy now realised the noise was actually coming from the flanking indirect paths – from a bay window, other walls and the sanded floors. We booked him in six months later to soundproof the flanking walls and treat the floor. Andy is now over the moon with the result, although he wished he had done it right first time, saving thousands of pounds.

CASE STUDY

Mark and Anita suffered from neighbour noise and wanted privacy. They first decided to soundproof the main bedroom in order to get undisturbed sleep, and were delighted. Their problem came when they decided it would be best to do the kitchen extension next instead of first soundproofing downstairs. The builder who did the extension offered to install some soundproofing to the party wall before putting in their dream kitchen.

Unfortunately, the soundproofing attempt by the builder amplified the noise from next door, and the new extension channelled the neighbour noise into the lounge. The neighbours would complain about noisy children and Mark and Anita were unable to entertain guests in their new kitchen because the noise would be transmitted not only next door but also up into the kids' bedrooms above.

After realising the mistake of getting the kitchen fitted first and also the true costs of having to uninstall it and remove the kitchen floor in order to solve the problem, Mark and Anita decided to live with their new kitchen renovation as it is.

To address the unwanted noise downstairs, my team soundproofed the front lounge instead of the kitchen, reconfiguring that as their family room to entertain and have people round. If they had installed specialist soundproofing first, the right materials could have been specified and installed before the new kitchen, and their beautiful open-plan kitchen would be their family entertainment room.

If you are not ready to invest in soundproofing because you have other projects, make sure you have a soundproofing expert's opinion on those projects, even if you think that the new work is not going to be affected by noise.

If your house suffers a noise problem, then it is highly likely the new addition to the house will have that noise problem too.

Get advice on materials so that the material you use does not amplify the noise. For example, aerated blocks, rigid insulation and modern plasterboard techniques all seem to make noise worse, so instruct the builders to use alternatives.

Some people are unaware that they don't spend much time at home, just avoid the noise, or accept, 'I live near a busy road' or 'It's a semi, that just how life is'... and learn to live with the noise in their house.

Some people think that having an attached house or living on a busy road means they will hear noise and it is just how life is. They convince themselves to live with it, so they spend a lot of time away from home, avoiding the noise. They don't have people to stay and may wear ear plugs to sleep. As a result, they don't enjoy spending time in their own home, feel stressed and embarrassed, and don't do the things they would like to. When they understand that the noise can be treated and invest in proper soundproofing, the results are life-changing – like moving into a new home.

CASE STUDY

Noel lives in a semi-detached house in Stratford-upon-Avon. He loves the house and the location, and likes to play music. He had a large house extension, renovated the kitchen and bathroom and created an office so he could work from home.

When a family moved in next door that made as much noise as Noel did, he found himself going to the gym seven days a week, sometimes twice in one day. He took several holidays, always buying nice things for his house, to make it feel more homely; but spent little time enjoying it. Noel failed to realise he was escaping his house.

When he realised the noise from the neighbours was disturbing his sleep, he started to see how the insulation of his house was affecting every part of his life, work and health. Due to the beautiful new kitchen renovation downstairs, he decided to have the bedroom soundproofed first so at least he could sleep.

He was astonished with how it changed his life; he started to watch films and even subscribed to Netflix. He was able to have a lie-in on his days off and spent more evenings at home.

Noel then decided to get the front guest bedroom soundproofed, which meant guests stayed over for the first time in ages. Then he had the office soundproofed, which allowed him to work from home; he could bring clients in, confidently knowing they were not going to be disturbed by noise from next door, and make professional calls without being disturbed. He now feels comfortable in his own home, has quit his gym membership and has equipment at home for exercise.

You might want to ask yourself these questions:

- If you could not hear your neighbours, would you spend more time at home?

- If you were 100% sure your neighbours could not hear you, would you spend more time at home? Would you play more music? Would you spend more time working from home, and more time with the family?

- Imagine that, after five weeks' renovation, your house could be as soundproof as a properly built detached house. Would you invite more friends round? Would you have more parties? What would you do with complete privacy?

Feel comfortable in your own home, be able to relax in the peace and quiet and do not let noise annoy you. There *are* solutions to most or all types of noise problem.

Summary

When looking to solve a problem of unwanted noise in your home, it can be easy to go down the wrong route, costing time and money and stress. Hopefully these case studies have given you an insight into why people soundproof their homes; maybe you can relate to a couple of them. Do any stand out as similar to your situation? Maybe you have tried soundproof wallpaper or had a builder in to soundproof the party wall? Or are you that unlucky reader that should have got soundproofing before renovating the kitchen? Maybe you believe your house is perfect. Maybe you have friends who tell you soundproofing is not effective and you are hesitant to go through the renovation.

In this life there are things that we do not control, which are not worth our time and energy – the weather, a virus pandemic or change in legislation. Noise entering your home, on the other hand, is something you *can* control; it *is* worth your time and energy. Now let me show you how…

PART TWO

THE FOUR-STEP SOUNDPROOFING METHOD FOR A QUIET LIFE

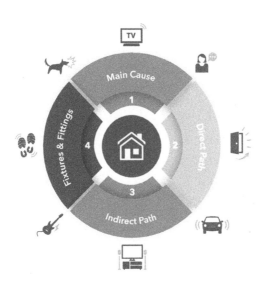

Step 1: Treat the main cause of the problem: Identify and fix things such as missing bricks, holes in the structure, poor renovation or thin walls, to be confident the underlying cause has been removed so you have a solid foundation to soundproof.

Step 2: Treat the direct noise paths: Apply a soundproofing system to the direct noise paths to barrier the noise transmitted via the separating ceiling, wall or floor, and reduce noise directly.

Step 3: Treat indirect noise paths: Identify and address indirect noise paths where the noise is coming around the separating structure, such as floors, ceilings, RSJs and window walls.

Step 4: Treat the fixtures and fittings: Install fixtures and fittings in a way that does not compromise the soundproofing.

4
Step 1: Treat The Main Cause Of The Problem

Here we are referring to something fundamentally wrong with the construction of your house, which lets noise in or out and/or amplifies the noise, making the problem worse. Every problem will have a main cause and it is very rare that the main cause is 'just thin walls between you and your neighbour'. I actually cannot remember the last project that was 'just thin walls'. More often than not, there are several main causes.

Addressing the main cause of the problem can include investigating a whole range of modifications to a house such as:

- A fabricated chimney stack made from plasterboard and stud wood

- The removal of a chimney stack

- Hollow, uninsulated RSJ running into the party wall

- Filling in the void below a suspended floor with concrete, without leaving an insulation gap between the concrete and the separating party wall

- A loft conversion using modern building techniques and materials that emphasise noise

- Uninformed attempt at soundproofing

Missing bricks

It may seem surprising how often this is overlooked. Bricks missing in the structure of the house create holes that allow noise to travel straight through. Bricks are often missing where floor joists (upstairs or downstairs) enter the separating party wall. Floor joists are the structural supporting wood that runs across the room; your floor boards are fixed to them. If your floor joists run into the party wall (you will be able to tell, because the floor boards will run parallel to it), then you may well have holes in the structure.

Also, bricks become loose over time. As people walk over the floors of your house, the wooden joists move, which eventually causes the bricks around them to become loose. Your neighbours have the same, as their floor joists also enter the party wall and their bricks may also have come loose over time. A house built in the 1920s, for example, would have had people walking all over the floor boards for the last 100 years. It is quite common to find holes in the party wall beneath the boards of a bedroom floor.

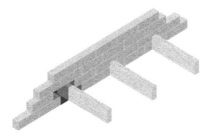

Missing bricks around floor joists

In one memorable project, we removed the bedroom floor boards to check for the main cause of the problem. Even though a double row of bricks created the separating wall between the two houses, we could see the neighbour's cream carpets and a dog's nose sniffing as it barked at us. The neighbour's dog must have thought there was a rat in the wall!

The issue of missing bricks also commonly occurs in the loft, where a separating wall meets the roofing felt. Any wooden joists, beams, purlins or RSJs sitting on the separating party wall could potentially have bricks missing around them, creating a clear passage for sound to transmit between the properties. These holes must be filled to prevent noise from the neighbours coming through the hole, into your loft and then down into your bedroom beneath. Bricking up these holes also helps maintain fire separation.

Occasionally we find bricks missing from a party wall in the suspended floor void for ventilation reasons. Our website shows some video and photographs of missing bricks and mistakes we have found: www. quietco.uk/blog/soundproofing-semi-detached.

Please do not cover up air bricks at the front of your house or you will have serious problems with damp and dry rot.

Separating wall in the loft

With bungalows and some old cottages or terraced properties, it's important to check that the party wall continues through to the roof. It's horrible to pop your head up and notice that it doesn't after you've already moved in, as this could be a difficult problem to sort out. You'll have to get it bricked up to get a decent amount of privacy, and get the neighbour involved too.

You'll need to check the construction of the roof, and understand how your roof breathes, to make sure the house was not built that way for a reason. The load-bearing capacity of the wall underneath needs to be calculated, especially if using dense 7.3 Newton blocks or laying bricks frog side up. To be absolutely sure you block this wall up correctly, I would recommend taking the advice of a roofing specialist who knows about condensation risks, as well as a structural engineer who can check the walls beneath and the potential additional load.

It is very common to find holes even in recent or newly built properties. This is a side effect of the culture of speed; noise through a hole through to next door is just not considered. Some builders believe these holes will never be seen; others cover them over with materials that do nothing for sound reduction.

Voids

If not insulated with acoustic mineral wool, the ceilings above bay windows channel not only road noise but also noise from neighbours.

Old cottages, pubs and hotels nearly always have hollow spaces that channel the noise between floors. These voids include boxing around toilet waste pipes, dumb waiters or kitchen lifts, decorative ceiling cladding and old chimney flues.

Newly built apartments and three-storey semi-detached new-builds often have plenty of voids, in hollow stud walls, uninsulated floor voids and, worst of all, boxed-in structural steel that runs the entire height of an apartment block. New properties also tend to have boxed in waste pipes. It is quite common for fire doors slamming, and even voices, on the ground floor to be audible on the top floor of a three-storey new-build apartment block. Like any wind instrument, the noise is louder at the other end. These voids must be insulated to stop them from channelling noise.

CASE STUDY

Mary lives in a semi-detached house built in 2010. She could hear her neighbours hoovering, kitchen pots, pans and microwave, an alarm clock; and because she could hear her neighbours while they were on the telephone, Mary felt she could only take phone calls in her conservatory.

She wanted the bedroom soundproofed first to minimise the disruption to her business, which she ran from home; later, she wanted the lounge soundproofed.

Understanding she would not get the full benefit upstairs until the lounge had been fixed, Mary had the bedroom soundproofed, removing plasterboard from the walls and creaky chipboard floors, too. Under the floor, we found a big void, boxing in a waste pipe, which channelled downstairs noise straight up into the bedroom. This boxing sat right on the party wall and ran downstairs between Mary's lounge and kitchen.

We knew we couldn't get to this boxing until we soundproofed the lounge. Mary was made aware the noise reduction would not be as expected in the bedroom until we could return to treat the main cause, the void.

Despite installing a premium floating floor and a ten-layered wall system in the bedroom, and treating walls and ceiling to reduce structural impact noises, noise from the neighbour's kitchen still woke Mary during the night.

Three months later, as scheduled around Mary's business activities, we returned to treat the lounge. To give her peace of mind, on the first day we opened up the waste pipe boxing and soundproofed that immediately. It was a relief for Mary to see the huge void channelling the neighbour noise to her bedroom and know she would get a massive reduction in noise in the bedroom from insulating it.

Every noise has a main cause and it is not always in the room to be treated. Each problem must be addressed on its own; rarely is there a benefit from treating just the party wall in every room. In Mary's case, treating just the bedroom gave her little reduction in noise.

Use of the wrong materials

Particular materials such as closed cell polyisocyanurate (PIR) board, a rigid foam sandwiched between aluminium foil used in roofs, floors and wall areas as insulation,[17] channel noise. If your house was built or extended in the last twenty years or so, you will most likely have this product in your house somewhere.

17 See, for example, www.celotex.co.uk/resources/faqs, accessed 15 October 2020

This material is great for thermal insulation but terrible for sound insulation.

We find that rigid, aluminium foil-backed insulation products seem to make both environmental and neighbour noise worse in homes. We always remove it and notice the difference immediately. We are not sure exactly why this is, as there has not been enough research published into these products and their sound insulation qualities. We do know, however, that sound travels very fast through aluminium, the foil coating on these products. In Steve and Natalie's barn conversion (see Chapter 3), the low-frequency noise from tractors and farm equipment travelled so easily through these products that it made the noise unbearable.

We have found that most foam-based products do very little for sound reduction. When you can achieve airtightness by spraying foam-based products, this will give a small perceived improvement with some high frequencies, but we have found that it is not at all effective for (low-frequency) road traffic noise and noise from neighbours. When the foam cures and solidifies, it bridges any isolation between different materials it is applied to and therefore makes structure-borne noises worse. Spraying foam into difficult-to-reach areas gives massive thermal advantages over many insulation products on the market, however.

'Dot and dab' – a possible main cause of the noise

I have mentioned 'dot and dab' a couple of times. The use of plasterboard fixed with dollops of plasterboard adhesive has been proven by the National House Building Council (NHBC) to increase noise (on average) by up to 10dB, a doubling in sound intensity as we hear it.[18]

'Dot and dab' involves fixing plasterboard directly to an area that needs plastering or making good. The plasterboards are cut to size, positioned and, when ready, stuck to the wall using strategically placed dollops of drywall adhesive to the surface to be treated. The boards are then levelled, creating a smooth surface ready for plastering. This method is now frequently used in new-build properties as it is very easy to carry out; most builders use this technique on renovation or extension projects. The ability to directly attach plasterboard to masonry gives a considerable speed advantage over traditional methods. It became established in England during the late 1970s and was developed to replace those methods (sand-and-cement render applied to the walls and smoothed off, ready for plastering before painting).

Traditional wet plastering is very messy; even the most skilled plasterers get messy applying sand-and-

---

18   See www.quietco.uk/blog/nhbc-report-on-dot-and-dab-plasterboard

cement render to a brick wall. Each coat of sand and cement takes about twenty-four hours to cure before the next coat can be applied. To give you an idea, an average lounge would therefore take one plasterer two or three days to plaster using this traditional technique.

'Dot and dab' involves mixing up a bag of plasterboard adhesive with a mixing paddle, so is also quite messy. However, a whole room could be 'dot and dabbed' in a day by builders, carpenters or any competent do-it-yourselfer, and a small room could have finishing plaster applied the same day. The speed at which a room, or even a whole house, can be made ready for finishing plaster has revolutionised the building industry. The cost is low, it is a much faster way of preparing a brick or block wall and the faster drying time means walls can be painted much sooner.

The problem with 'dot and dab' techniques is that, although you are putting an extra layer on the wall, you are amplifying the noise from your neighbours. A cavity is formed between the dots of adhesive, which creates a drum effect. A site investigation by NHBC acousticians revealed that removing 'dot and dab' adhesive and exposing the blockwork behind, improved the sound insulation of the party wall and separating floor. The single-figure performance improvement was 3dB; however, at the frequencies where the human ear is most sensitive (1,000hz to 5,000hz) sound insulation improved by nearly 10dB

– just from removing 'dot and dab' plasterboard. My own acoustic engineers managed to match the graph in the NHBC report exactly.

What does this mean for you?

If you have a floor and ceiling that is soundproofed to meet Part E of the Building Regulations but your walls in that room are then plasterboarded using the 'dot and dab' technique, then your walls are making the noise from your neighbours worse. Simply removing the plasterboard will improve sound insulation. A separating floor/ceiling without 'dot and dab' plasterboard on the wall will perform better than one with it. Not only will it make the noise worse but it will channel the noise across the whole of your house. The pockets of air behind the plasterboard transmit the noise into other rooms. This is most obvious for flat renovations and new-build apartments, where noise will come through party walls and resonate between the plasterboard and the structure it has been applied to.

'Dot and dab' can be one of the main causes of increased road traffic noise in newly renovated houses, as the cavities behind the plasterboard can go all around the room, channelling road noise from the window. Noise can also be channelled from the loft (eg aircraft noise) and between floors.

CASE STUDY

Bob lives in a semi-detached house in Cheltenham. He is a professional tradesman with an eye for detail. Every Friday and Saturday night the neighbours decided to sing along to the TV or get the karaoke machine out. This made life uncomfortable for Bob and his son. Bob thought he could solve the problem himself by soundproofing the bedroom. He got his colleagues to help install sound-resistant plasterboard and some heavily marketed soundproofing materials to all the walls in the main bedroom. Bob also fitted new skirting and architrave, painted and decorated, and fitted new carpet and a load of new gadgets.

After moving back in, he realised immediately that the noise sounded twice as loud as before; he could now make out words and follow the songs on the karaoke machine. He also heard many new noises he had not heard before. He began to lose sleep and the whole ordeal affected his work and family relationship.

After a consultation and understanding how he had made the noise worse, Bob decided to hire us to cure the main bedroom. He removed the 'dot and dab' 'soundproofing' from the bedroom walls ahead of the new installation. We treated the floor, ceiling and walls and Bob no longer hears the neighbours in his bedroom.

He can go to sleep when he wants and wake up when he wants, and Friday and Saturday nights are no longer a problem.

If you do it yourself, I suggest you:

- Do not use 'dot and dab' techniques

- Do not go over the top of any existing 'dot and dab' plasterboard; it must be removed

- Do not scrimp on quality; you will be pleased you didn't after you move back into the room

If in doubt, pay for some soundproofing advice – a consultation from an installer that does soundproofing work all the time. A few hundred pounds spent on advice might be considerably cheaper than having to take a failed installation down after painting and decorating.

For any attached property, whether a flat, apartment, semi-detached house or terraced property, if one person renovates using 'dot and dab' the neighbours may then be able to hear one another. Some argue the problem lies with the fixing method: 20% of the plasterboard should be covered by the dab adhesive. Having removed this 'dot and dab' technique from domestic homes every month for the last decade, I am convinced that *any* use of this technique will make noise worse, regardless of how much adhesive is applied.

What is the alternative? Traditional sand-and-cement render is much better for sound because it adds an airtight mass to the wall. Rendering internal brick and block walls has been used since the Romans and is still found in buildings built before the middle of the twentieth century. It turns out that it is better for sound insulation, and more environmentally friendly than 'dot and dab'. Unfortunately, this rendering technique is not widely used as it is more expensive and the number of qualified plasterers capable of levelling a wall with sand-and-cement render has massively reduced over the last twenty years. Most plasterers will want to use gypsum backing plaster, which is better than 'dot and dab' but does not have as much mass as a sand-and-cement render.

The best approach is to install a soundproofing system that will deaden the problematic wall. Use a combination of rubbers, highly flexible (visco-elasticity) materials and dense board that will not resonate in the wall and isolate the new wall surface from the structure. This should use up little space after you have removed the original plasterboard.

Aerated blocks as a main cause of noise

Aerated blocks need to be mentioned here because most modern houses are built using them and they contribute to sound transmission up and down walls; low-frequency noise seems to absolutely love aerated blocks. Using them on an extension could mean you will have to soundproof a wall that is not even touching the neighbour's wall (see Chapter 5).

They are manufactured with sand, aluminium powder, calcined gypsum, lime, cement, water and by injecting bubbles into the mix to reduce the density. This concrete mix sets to form a cellular porous concrete foam block, which is great for structure, thermal insulation, fire and mould resistance, and sound transmission too.[19] They are grey or white, lightweight blocks and can sometimes be recognised by the scratches on their face. Even if you apply render to these blocks, general modern living noise will transmit up and down and through them very easily. TV noise, conversation and particular tones of voice and dog barking go straight through them.

Any modern apartment with internal walls constructed using these blocks will be expensive to soundproof because you will have to isolate your finished walls from these blocks.

If you live in a recently built property and you experience noise from neighbours, you might say, 'It's as though my neighbours are in my house'. This phrase is often enough to tell me your house or apartment is constructed using aerated blocks with 'dot and dab' plasterboard fixed to them. Some clients who did not notice their neighbours before an extension or hear road or traffic noise have found the noise so bad they have had to move.

---

19 Sahu, MK and Singh, L, 'Critical review on types of bricks type 3: AAC block', *International Journal of Mechanical and Production Engineering* 5, 11 (2017)

CASE STUDY

'It's not coming from the new extension... that's not even attached!'

Peter and Caroline live in a 1930s semi-detached house with a single-storey rear extension. They started to hear the neighbours' music, TV and, worst of all, the French doors banging every time the neighbours went out to the back garden. For the first year or so, they were convinced it was the new stereo equipment, change in neighbour habits or the children getting older and making more noise. We began soundproofing the back lounge/dining room, bricking up the holes in the structure where I was able get my arm through to next door beneath the floor boards.[20]

We disconnected the floor joists, reducing transmission of impact noise getting into the structure, and we installed sound insulation to the direct noise path and some of the indirect noise paths. Peter and Caroline had only Steps 1, 2 and part of Step 3 done. It then became obvious to them that the new extension was an indirect noise path (see Chapter 6).

We returned to Peter and Caroline's house and uncovered 'dot and dab' plasterboard, aerated blocks and aluminium-foiled rigid insulation used in the sloping roof. We removed the plasterboard, deadened the blocks using our trademark 50mm wall system and removed the rigid insulation. Despite the longest soundproofing project ever (because of the lockdown), Peter and Caroline are delighted with the result.

---

20 See Quietco videos, www.quietco.uk/videos

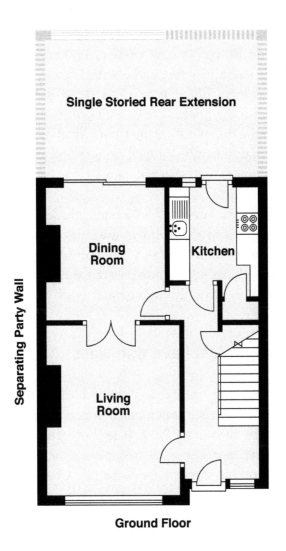

Floor plan

Building an extension or conversion before giving priority to soundproofing can lead to amplifying the noise in your home. Making it right could involve gutting

back to blocks and bare wooden rafters, removing fixtures and fittings (a new kitchen, say) and, together
with the soundproofing and disruption, the cost is too
much for many homeowners and they decide to move
away from the noise. If only the right materials were
used in the first place, or the sound insulation properties of materials was as well understood in building
regulations as their thermal value.

I suggest you specify 7.3 Newton blocks. They are
very heavy and the builders are unable to build as
quickly as with aerated or ultra-light blocks; but this
dense block has great noise resistance and noise finds
it difficult to travel through greater mass.

You might just have thin walls

Before you get too excited about finding the main
cause of the noise, remember this is only one of the
four steps in our method. It is, however, the most
important, because most people go straight into
soundproofing direct noise paths (see Chapter 5) and
miss the main cause, then wonder why their system
has not reduced the noise. Take the time to find the
main cause of your problem.

Summary

Step 1 of the four-step soundproofing method: treat the
main cause. This is something that is fundamentally

wrong with the construction of the house. Holes in the structure, the materials used, poor building techniques, and voids left in the structure are the main offenders.

If you can make out words, laughter etc, or hear music clearly, then it is highly likely you have holes in the structure. High-frequency sound is a small sound wave; it readily passes through holes and is easy to barrier with most bricks and mortar.

The 'dot and dab' technique is used in nearly every modern home and renovation, but is making the noise worse for you. Do not use it on any extension or alteration or renovation; use wet plastering, insulate between battens or seek soundproofing advice. Also, do not use either rigid insulation or aerated blocks. Request the architect and builder to specify alternatives.

If your builder says they know how to soundproof or will soundproof your extension, check they are not just insulating a stud frame or 'dot and dabbing' 'soundproof' plasterboard. Ask for an estimated percentage noise reduction from the system they are applying or speak to other people who have had soundproofing installed by this builder. If in doubt, consult a specialist soundproofing installer for advice.

5
Step 2: Treat The Direct Noise Paths

After using many soundproofing materials sup-pliers over the years, it has become increasingly frustrating to see them creating wall, floor and ceiling systems that could easily be distributed on a single pallet, and are designed only to treat the direct noise paths. Products sold and marketed are designed for a slight noise reduction, ease of installation and dis-tribution. They focus on treating only the separating party wall between homes, with little consideration for indirect noise paths (see Chapter 6). These systems often have highly inflated decibel values attached to them with a claim that these figures are laboratory-tested and the caveat that they must be installed as per manufacturer's instructions to achieve the claimed figures. They can say (and claim) whatever they want because the market is not regulated and who really knows what 60dB sounds like on one type of construction compared with another?

Over the years, as new materials became available and distribution networks from other countries massively improved, we continued to try out new technology, enhancing the combination of acoustic products and techniques that we use. We test our systems in actual homes rather than a laboratory, so we know how they will reduce domestic noise in people's homes. From this we have developed our own systems (like the floating frame) to treat direct noise paths effectively. We have managed to keep the floating frame technique a secret all this time and I will be sharing the engineering brilliance of it a bit further on.

Direct noise paths are paths along which noise transmits directly, such as a separating party wall, alcoves either side of a chimney breast, or the fire back (area behind your fire) in a chimney stack. In a bedroom the direct noise path for road traffic noise would be the windows and the walls around them. In a flat or apartment, direct noise paths can be the floor or ceiling, depending on whether the neighbour noise is coming from above, or below, or both. A terraced property will have direct noise paths on each of the separating party walls.

Walls

There are many different ways to soundproof a wall, with many materials suppliers offering different systems. Having installed soundproofing to domestic homes for well over a decade, I want to share the system that we use most often today.

What is the best system to install on a party wall?

The floating frame is a a component in two of our most popular wall systems, which we have kept secret all this time. We have watched many new soundproofing companies enter the market and promote their illustrations and wall systems. All of them use traditional building methods, such as aluminium stud partitions or the usual Canadian Lumber Standard (CLS) stud frame method, but none come close to the floating frame. We have found that 30mm panel systems installed to the direct noise path rarely work in a domestic environment.

Now trademarked with design rights, our floating frame is installed against the wall on any direct noise path that needs an exceptional barrier. It is an independent frame isolated by a 25mm gap from the ceiling, walls, floor and the party wall. It is constructed under pressure so, before we install any soundproofing materials, you have a resilient frame isolated from the structure, which noise will find very difficult to get into. We then install a number of layers of high-quality specialist soundproofing materials, which can include visco-elastic membranes, rubber, cement board, barrier mats with different densities, fibreboard (like the mat under your bunsen burner at school, not the asbestos type), resilient bars, sound-bloc board and acoustic mineral wool.

Why is it effective?

Our installation is designed to barrier a wide range of frequencies. Acoustic mineral wool (we use a dense product, 100mm thick weighing $60kg/m^3$) gives the system superior absorption of mid- to high-frequency sound transmission and cement board reflects that sound back into it. Fibreboard and barrier matting give the system stiffness, which reduces transmission of the really low frequencies. The materials, all interlocking and installed on the floating frame using our unique, tried and tested technique, work in combination and, crucially, the frame is isolated from the house structure and not just from the party wall. The combination of products barriers the noise directly but also reduces the resonance of the walls. For the considerate noise maker, the floating frame prevents your noise transmitting through the house structure into other rooms and into the neighbour's property.

Sound waves travel more quickly through many materials than they do through air; in fact, sound waves tend to lose less energy through many construction materials than when they travel through air:[21]

Through (material)	Speed of transmission (m/second)
Rubber	60
Air at 40°C	355
Air at 20°C	343
Lead	1,210
Sand	1,625
Polystyrene	2,400
Common red brick	3,600
Softwood (pine)	3,320
Hardwood	3,960
Zinc	4,200
Glass	4,500
Steel	5,200
Aluminium	6,350

We construct our floating frames using wood rather than the aluminium stud partitions that are popular in the building industry, and the table shows why.

---

21 See www.britannica.com/science/sound-physics/Circular-and-spherical-waves for how sound waves propagate and move, and 'The speed of sound in other materials' by EP Ortleb and R Cadice (1993), www.nde-ed.org/EducationResources/HighSchool/Sound/speedinmaterials.htm, accessed 15 October 2020

Alcoves

Many people I speak to about neighbour noise on a party wall are convinced the noise is coming from the alcoves either side of the chimney stack. Many clients email with dimensions of the alcoves and just want a wall system installed to directly barrier the neighbours' TV, music and conversation. The reality is that this will give just a small reduction in noise, and only when you stand directly in front of the new soundproofing. Once this direct noise barrier has been installed, it will it become obvious that indirect noise paths also need addressing (see Chapter 6).

The thinnest system we install on party wall alcoves is 50mm. This will considerably reduce TV and conversation noises, but the occasional shout will still come through. Of course, a window or door might prevent you from installing a thicker system. One of the biggest concerns for people thinking of having soundproofing installed to the direct noise paths is loss of space. The chances are that after painting and decorating you will not notice the space loss, unless you have a particular piece of furniture fitted to your alcoves.

A 95mm wall system will barrier louder noises such as music, a baby crying, gatherings of people, gaming or movies. Our 95mm system achieved a sound insulation value of 63dB DnT,w + Ctr (see Appendix for what this term means) on the alcoves of a 1930s semi-detached house. All the flanking areas were treated, and floor joists were disconnected, as part of the installation (see Chapter 6).

For more extreme cases – neighbours that are hard of hearing, bass music, grand pianos, acoustic drums, dogs barking – and structure-borne noises like extractor fans, boilers, kitchen-cupboard or wardrobe doors installed on the other side of a separating party wall, I recommend installing a 150mm or even 185mm wall system to alcoves and the fire back.

CASE STUDY

Mark lived on his own in a two-bedroom, semi-detached, modern house. He was suffering from noise from the neighbours, including music, TV, conversation and children running up and down the stairs late into the evening. We installed our floating frame to the direct noise path, and treated the ceiling, window wall, bedroom floor above and isolated the fixtures and fittings (see Chapter 7). Mark was delighted: he no longer hears the music, TV, conversation and, best of all, no longer hears the children next door running up and down the stairs.

Fire back on the party wall

The area at the back of the chimney, behind your gas fire, log burner or electric fire, is a direct noise path. This fire back area is a very important part of Step 2. It may seem frustrating if you have a beautiful fireplace, perhaps the original from when the house was built… but it is probably not soundproofed! Behind that feature is just brick; depending on the age of the property there could be a single layer of bricks between you and your neighbours. What is the point of treating the alcoves and leaving this great big hole in the wall for noise to transmit directly through?

Any wall system that is applied to this fire back or chimney stack must have a good fire rating; products that go to 1,000°C exceed all current regulations and are available from most builders' merchants.

Most soundproofing installations to a chimney stack need to be carried out before any wood burner or gas fire is installed, otherwise these will have to be taken out and reinstalled after the soundproofing is in place. The different types of room heater, together with Approved Documents F, J, L, G, P of the Building Regulations, is a whole book by itself. Any gas fire, wood- or biomass-burning stove or open fire will fall under Document J, and removing it then reinstalling after soundproofing requires a building regulations application. The key thing to note is that you cannot have any combustible material within 200mm of the flue.

One client found their wood-burning stove was not the best choice for their soundproofed front room. Soundproofing is a good thermal insulator; heat finds it very difficult to escape a fully soundproofed room. Their stove heated the room so quickly, they found themselves trying to put it out before getting that second log on the fire. A stove expert would say they simply had the wrong stove for the room. Sound insulation really does keep the heat in, and many clients find an added benefit is improved thermal efficiency and lower bills. A small radiator only needs a quick blast and the room stays warm for ages.

Removing a chimney stack

Removing a chimney stack does reduce the mass on the separating party wall and therefore it is a good idea to install a soundproofing system across the entire wall in its place, especially in case your neighbours decide to remove their chimney stack in the future. When you have removed a chimney stack, you are often left with direct noise paths through holes in the structure, which must be filled with sand and cement before the installation of any wall system. The jagged spikey brickwork will need to be smoothed off in order to fit soundproofing. Removing the chimney stack and installing a 150mm floating frame with interlocking layers across the entire direct noise path can be better than treating the alcoves, chimney stack and fire back as individual sections along the separating party wall. It will also open up the room and give you more space.

Windows and window walls

The direct noise path for road traffic noise would be primarily through the windows and the window walls directly facing the road outside your home. However, sometimes rooms take a different shape to normal.

CASE STUDY

Jason lives in a 1½-storey modern detached house on a busy road. The sloping roof forms the walls and ceiling of the upstairs bedroom: one side has a sloping roof facing the busy road, which is the direct noise path along with the windows installed in this roof.

The house was perfect for Jason's needs, but road noise was costing him sleep. He tried window barriers, acoustic curtains, acoustic foams and white noise machines, to no avail: once he was woken up by a noise, he struggled to get back to sleep. He also found it difficult to concentrate while working from home.

He first updated the roof windows to new, acoustic-laminated versions with electric roller shutters, to see what difference they would make before committing to any major renovation to his bedroom. As expected, the new units alone did not reduce noise much at all.

I was able to show Jason how and where the road noise was entering his main bedroom. We applied the four-step soundproofing method to his main bedroom. In Step 1, we addressed the main cause of the problem by removing aluminium-coated rigid insulation that was channelling the noise from the road into the main bedroom.

In Step 2, we addressed the other direct noise paths. The roofing directly facing the road was treated with a ceiling system that took up 50mm space, but keeps the roof ventilated and uses vapour barriers that exceed building regulation requirements.

In Step 3, we treated indirect noise from the ceiling the other side of the room, a few hollow stud walls and the bedroom floor, and in Step 4 , we addressed the installation of fixtures and fittings (a built-in wardrobe, fitted furniture, air conditioning, finished flooring and secondary glazing).[22]

The reduction in noise after plastering was significantly more than the 50% reduction estimated before the project began: more like a 70% reduction. Adding soft furnishing, such as carpet, the bed and other furniture, reduced the road noise further. After secondary glazing was installed, Jason was able to look out the window, see the cars go past but not hear them.

Some people (and their pets) are disturbed by seasonal noise like fireworks. Upgrading windows and the window wall can help barrier that noise directly.

Most window walls facing a road or conflicting noise that need to be soundproofed are either old 9-inch-brick walls with no cavity, or built using lightweight aerated blocks. A barn conversion along a main road may have many window walls as direct noise paths: gables at one or both ends facing traffic, or a main

---

22 See the before and after video at www.quietco.uk/environmental-noise

wall backing on to the road, which will need sound deadening material applied and then be isolated from the structure, with 50mm to 150mm space loss.

Unlike walls, roofs and ceilings that can reflect *or* absorb sound waves, glass will allow high-frequency sounds and vibrations to pass through. This is why windows are generally the weakest point of insulation against environmental noise in the home. I get asked time and time again whether to get triple glazing installed to reduce road noise. It is the acoustic-laminated glazing that makes the difference, so long as the inner pane is laminated. By all means get triple glazing if the inner pane of glass is laminated and the unit is designed to reduce noise. Double-glazed, laminated acoustic glass with the right airgaps between panes will outperform *standard* triple glazing. The standard transmission class (STC) rating of the window should be in the region of 44dB. We have always used Velux double-glazed roof windows with a laminated inner pane. It may seem obvious, but not much improvement will be achieved if you upgrade to triple glazing with acoustic-laminated glass and still have large trickle vents installed to your new windows.

Ultimately it comes down to the quality of the window installation, the measuring-up of the windows and the workmanship of the team installing the windows. So what should you look for?

Most installers of windows will have the window manufactured smaller than the brickwork opening size by as much as 5–10mm all around the window. This allows the window installation teams to slot each new window in without worrying whether the window will fit, and having to plane down the frame or chop out brickwork. Using spacers, silicone, expanding foam and plastic beading to fill and hide the gaps, some companies can do a whole house in a day. Unfortunately, this is terrible for noise and you the customer will not get the laboratory figures for noise reduction quoted in the brochure for that window. The gaps let the noise in.

You might ask, 'How do I know if I have gaps?' You do. You can fill them with acoustic sealant or, better, add sound insulation to the window reveals (the window frame, of which the bottom forms the window sill). The window reveals are another key component in making sure you get the maximum integrity from your new window investment. Make sure the window reveals are not plasterboard, which channels the noise from these gaps all around the room. If they are, then remove the plasterboard and replace with wet plaster or (better) a soundproofing panel.

For maximum reduction in noise from outside, add secondary glazing to the window reveals. Secondary glazing is a separate, slimline window installed on the inside, normally within the window reveals. The secondary unit provides two layers of window glass isolated from your existing window. Secondary glazing using acoustic-laminated glass would provide the best sound insulation on windows against direct noise paths for many households, especially those situated around airports, railway lines, under flight paths, near industrial units or next to busy roads. Correctly installed, secondary glazing can barrier neighbour noise such as dogs barking late at night, and if you work shifts and need to sleep during the day they are great for reducing outside noise. Secondary glazing provides better sound insulation than conventional double glazing alone because it creates a larger air gap between the window and the new unit (as the table above showed, air transmits sound less rapidly than glass).

Please note that once you barrier the noise transmitted directly from outside, you will make noises transmitted indirectly, from the party wall, ceiling or floor, easier to hear (see Chapter 6).

Ceilings

If you have unwanted noise coming from an apartment or flat above you, the direct noise path is the ceiling. For ceilings, we use a system similar to the floating frame, isolated from the conflicting ceiling and surrounding walls in the room. Like the wall systems, the space available depends on the ceiling system used, and this in turn affects how much reduction in noise can be achieved. The type of noise is also very important to consider when treating a ceiling directly.

As you may have gathered by now, treating impact noise at source is massively more effective, ie treating creaky floor boards by screwing them down and isolating them from the walls is more effective than trying to fix a barrier to the ceiling below. Soundproofing impact noise such as furniture moving, footsteps, doors banging, cupboard doors closing etc is difficult from below. It normally requires 150–250mm loss of room height. Best results have come in older properties that have huge wooden joists and thick lath-and-plaster ceilings and therefore less structure-borne flanking noise down the walls. These older properties can also have higher ceilings that can take the kind of height loss required to reduce impact noises from above.

Our 90mm ceiling system installed to the direct noise path with seven or more layers of different types of acoustic insulation will offer 65% reduction in impact noise from above, and over 75% human-perceived reduction in airborne noise. To achieve more than this, a thicker ceiling system needs to be installed.

Floors

The most effective way of treating noise from a flat below is to install a resilient layer to the ceilings below your floor. However, this requires access to the neighbour's flat, which only happens with strong neighbourly relationships. Therefore, for most home-owners soundproofing the floor is the best alternative.

You may have tried thick carpet underlays, or even acoustic rubber underlays from specialist soundproof-ing suppliers. If the separating floor is wooden, these products will give a small reduction in noise if placed directly over the hollow uninsulated floor. If you pull up the floor boards and insulate the entire floor with acoustic mineral wool, then install heavy-duty acous-tic underlay on top of the refitted floor boards you will get a noticeable reduction in noise from below. The neighbour below you will benefit most with a reduction in footfall impact noise and airborne noise from above, too.

A floating floor is not glued or screwed to the joists or subfloor but made up of interlocking panels or sections of flooring glued together and resting on a resilient layer such as a foam or underlay. A floating floor can have as few as two layers (a foam underlay and a hardwood floor, say), but the best will include different types of rubbers, resilient foams and a solid reflective mass, which moves with the finished floor and converts the impact vibration energy into heat. It is very important that a floating floor is isolated from the walls all around the room and not glued or screwed to the joists. This isolation reduces transmis-sion of a wide range of frequencies that come from general living. The best results come from a combination of different acoustic materials installed between the floor joists as well as a floating floor, before the finished floor covering. Space loss for floating floors ranges from 11mm to over 100mm, with our most popular floating floor at 20mm offering 80% reduction in the impact noise from the conflicting floor to the room below. Floating floors also prevent structure-borne noise transmitting into your finished floor, vibrating the bed and waking you up, for example.

The most common noise problem in flats is impact noise from footfall and doors banging. To reduce impact noises from the flat above, you need to install a resilient layer below your neighbour's finished floor. When installing a floating floor to the floor above, fixtures and fittings such as kitchen, bathrooms, skirting, architrave, doors, and finished floors will all have to be uninstalled and refitted (see Chapter 7). These costs, and the disruption that comes with the work, need to be taken into account when trying to persuade your neighbour above to install a floating floor.

In my own house, I found floating floors effective at reducing the indirect noise from neighbours, but most satisfying was reducing the direct noise of the children jumping around upstairs in their bedroom. Before installing a floating floor, I would watch the ceiling light shake in the living room every bedtime.

With floors, it is important to make full use of the voids between the floor joists, installing a mixture of different acoustic materials to tackle different frequencies of noise before isolating the finished floor with a floating floor. This will barrier noise from downstairs directly.

Building regulations

Meeting building regulations for separating floors (Document E) is easily achieved by installing resilient layers both on top of the joists and, for best results, below them as well. This could help a developer meet the 'change of use' regulations, such as converting a terraced property into flats or an HMO (house in multiple occupation), or an old factory building into luxury apartments.[23]

There are a number of ways of achieving and meeting these building regulations and every project is different. If you are 'changing use', then you may be able to use what is already in place to your advantage, with different-size joists, concrete floors, stairwells, and service voids. Whether it is, basement conversions, shops, offices, restaurants, they all require the same four steps outlined in this book.

---

23 See footnote 5 for the different requirements for airborne and impact noise

Wherever possible, as well as a floor system, a ceiling system should also be installed to the ceiling below. Treating both the ceiling and floor separating two flats is what is required to completely barrier direct noise. This is not always possible and there are ways of meeting the regulations by carrying out work above without affecting the flat below; or doing all the work below without disturbing the flat above. Why disturb both flats when you can keep all the disruption in one, you might ask? Meeting the regulations on one side of a floor separating flats can be complicated, with the need for expensive, specialist acoustic materials. It can take more time than doing both flats. There are many factors to take into consideration, such as access, loads to walls and ceilings, loss of space and different types of fixtures and fittings (see Chapter 7).

Aircraft noise

There is plenty of evidence that aircraft noise affects cardiovascular health, disturbs sleep, affects psychological well-being and has effects on children's cognition and learning.[24] 'The study (Hansell et al., 2013) concluded that high levels of aircraft noise were associated with increased risks of stroke, coronary heart disease and cardiovascular disease for

---

24 See Clark, Dr C, 'Aircraft noise effects on health', Queen Mary University of London (2015), https://assets.publishing.service. gov.uk/government/uploads/system/uploads/attachment_data/ file/446311/noise-aircraft-noise-effects-on-health.pdf, accessed 17 December 2020

both hospital admissions and mortality in areas near Heathrow airport.'[25]

The direct noise path for aircraft noise is the roof of your house. At the moment, it may have nothing between the roof rafters (rafters are structural timber or steel used as part of the roof), so in the loft you will see roofing felt or tiles; or the materials I mentioned in Chapter 4 that make noise worse in your home.

To prevent aircraft noise entering bedroom ceilings, acoustic mineral wool must be installed between the rafters, while still allowing the roof to breathe, and a resilient layer installed to decouple each individual bedroom ceiling from the direct noise path. This may involve removing what is already in place and replacing it with good-quality sound insulation. The acoustic wool absorbs noise and forces it to change direction and the resilient layer(s) transfer the sound energy into heat energy, decoupling the ceiling deals with lower frequencies and reduces noises that vibrate through the structure and wake you up.

---

25 See Clark, Dr C, 'Aircraft noise effects on health', Queen Mary University of London (2015), https://assets.publishing.service. gov.uk/government/uploads/system/uploads/attachment_data/ file/446311/noise-aircraft-noise-effects-on-health.pdf, accessed 17 December 2020

Summary

Now you understand Step 2, treating the direct noise paths. This is a key step, but not necessarily the most important.

Panel systems rarely work for the direct noise path alone; they should only be used where space is at an absolute premium or to treat flanking walls (see Chapter 6). As a barrier to extreme direct noise, use an independent wall system or the floating frame on party walls, alcoves, fire backs and any wall through which noise can come directly.

Applying soundproof insulation to the wall at the back of the fire (fire back) on a separating party wall is very important, to prevent unwanted noise transmitting directly.

To barrier road traffic and environmental noise, use double- or triple-glazed windows with laminated acoustic glass for best results, or upgrade the windows with secondary glazing, getting the right air gap between the panes of glass.

For direct noise from below, use a floating floor.

When treating a ceiling, you are usually only installing half a system; the other half ideally needs to be fixed to the floor above. For direct noise through a ceiling, use a suspended ceiling isolated from the existing ceiling and walls.

My company has invested heavily, on our own and in partnership with organisations and universities, in research and development into the optimum combination of different materials to effectively treat direct noise paths. We have also invested in finding non-combusting, environmentally friendly, sustainable alternatives, to align our company with its vision and values and UN Global Goals 7 and 12. See more at www.quietco.uk.

6
Step 3: Treat The Indirect Noise Paths

If you were only going to read one chapter in this book, this is the one. Why do so many people go to great lengths to soundproof only to have it make very little difference? The answer is simple: flanking transmission, otherwise known as indirect noise paths.

If flanking transmission is not properly addressed in a soundproofing installation, the project will fail. Since indirect noise paths are sometimes louder than direct ones, leaving them untreated will mean that noise is as troublesome as ever. In this chapter, we'll be looking more closely at indirect noise transmission, and how to correctly soundproof across a range of structures.

Why most soundproofing attempts fail

Picture this: you are in paradise, sitting on a beautiful, sandy beach, admiring the view, the tide is coming in and the water is getting closer. How are you going to stop your feet from getting wet? You might decide to construct a simple barrier made of sand running along the bottom of your feet. Problem solved, right?

You quickly notice that, although your barrier stops a small amount of water from reaching you from straight-on, the tide starts flowing round the sides of your barrier. 'OK,' you think, 'time to build a sand barrier around the sides as well.' You then notice the water can flow over both your front and side barriers, once it gets deep enough. Just as you're about to build a barrier to cover your feet, you also realise that the water's getting in beneath all your barriers, and seeping in indirectly. You get the picture. Your home is the peaceful paradise, and the sound waves are the water creeping in from all directions – around, over, under and right through any barriers you build. Just as building a single barrier between you and the ocean is not going to stop the rising tide, simply putting materials on top of your party wall is not going to stop all the sound that travels along your flanking walls (indirect noise paths) and through your entire home.

Treating the flanking areas is key to any successful soundproofing project. I realised this early on when treating my own house. My company still has the fundamental philosophy of treating all the flanking areas, knowing that doing so means our quotes and estimates are twice if not three times as expensive as our competitors' quotes. However, we know these are essential in ensuring our customers get the best from their soundproofing investment and actually reduce unwanted noise significantly.

What are flanking areas and why are they important?

'In buildings with traditional heavy constructions of concrete or masonry, about 50% of the sound transmission between rooms with a common dividing partition is through flanking transmission.'[26]

If flanking transmission accounts for 50% of the noise, soundproofing the direct noise paths alone will give you a lot less than 50% reduction in noise. Even if you have a thick concrete floor between you and your neighbour below, you will still hear that TV, conversation and music through the concrete structure, though more muffled. This muffled noise is the sound radiating from the concrete surfaces. Your neighbour's music that starts off as airborne noise carried into their walls, ceilings and floors transmits through the building structure and then radiates out of the walls, ceiling and floors in your room as an airborne noise. The volume of the source noise will determine how

---

26   See Rindel, JH, *Sound Insulation in Buildings* (CRC Press, 2018)

clear the noise you hear is. For example, if someone's shouting you will probably make out words; if music's loud you might guess the music track. This is because when noise reaches a certain volume (decibel), it will enter the structure of the building.

The figure below shows how noise in an upstairs room transmits through the property and through the party wall and flanking areas. In many cases, the flanking paths (light grey arrows) are louder than the direct transmission paths (dark arrows). This means that you won't get the full benefit of soundproofing the upstairs rooms until you've also treated the down-stairs rooms and addressed the flanking areas.

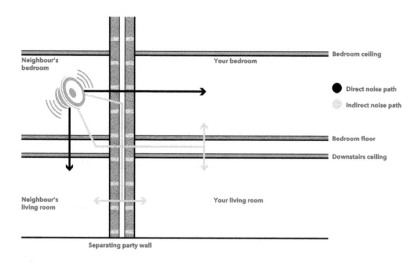

Upstairs bedroom noise. Adapted from Rindel, JH, Sound Insulation in Buildings (CRC Press, 2018)

Indirect noise paths are complex and numerous. Around a party wall, for example, the floor, ceiling, window wall and any wall adjoining the party wall could be an indirect noise path. Also, an uninsulated RSJ might be connected to that wall, or a bay window ceiling, chimney stack, stud wall, cupboard under the stairs, *en suite* bathroom or wall separating front room from back room.

The walls joining a ceiling or flat floor could be indirect noise paths. As I showed in Chapter 1, 'dot and dab' plasterboarded walls amplify noise from above or below. 'Dot and dab' plasterboard fixed to the walls below or above a ceiling radiates more unwanted noise than the direct noise path through that ceiling.

This is why indirect noise paths can be understood as the 'weakest link' in any soundproofing project. In many homes, the unwanted noise is louder from these indirect noise paths than from the direct noise paths.

Though treating flanking transmission may seem like a lot of work and expense at the time, identifying these areas and ensuring they have adequate sound insulation makes a huge difference to any soundproofing investment.

Flanking areas that must be addressed

Ceilings and floor voids

When soundproofing the direct noise path to a wall, the system should always be tucked down below the floor, and the floor should then be insulated with a combination of acoustic insulation materials. This prevents the noise from passing *under* any wall system. The ceilings are equally important: most ceilings on the ground floor of attached houses (semi-detached or terrace properties) will have uninsulated ceiling voids.[27] If you tap them with your hand, they sound like a drum. These uninsulated voids are a key flanking area, which must be soundproofed to prevent unwanted noise passing *over* any wall system.

Some semi-detached properties have a suspended floor void beneath the ground floor, which can often be a main route for noise. The figure opposite shows flanking indirect noise (grey arrows) through ceilings and suspended floor voids. It is crucial to insulate these suspended floor voids. Please note the floor must be insulated in a way that allows the wooden joists and floor to breathe and you must not block up vents or air bricks, otherwise you could cause damage to your property requiring major repairs to your house.

---

27   Detached houses probably do, too!

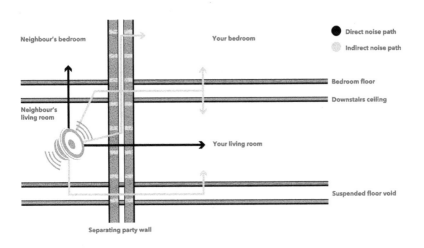

*Flanking noise through ceiling and suspended floors.
Adapted from Rindel, JH, Sound Insulation in Buildings
(CRC Press, 2018)*

Window walls

Another very common flanking indirect path for neighbour noise is any window wall joining the separating party wall. To find out whether your window wall needs sound deadening, simply put your ear on it, finger in the other ear and listen. The window wall section and reveal closest to the separating party wall will most likely be transmitting some indirect noise. If you hear the noise, don't ignore it. This is neighbour noise radiating out of this area into your room and could be a key area to treat as part of your soundproofing investment.

CASE STUDY

Roger and Alice's neighbours were hard of hearing and elderly. They enjoyed watching TV all day, incredibly loud, and being hard of hearing they naturally shouted to one another across the house. Our team had the honour of installing our trademarked floating frame in their home, but when we came back the next day to plaster the soundproofing system on the party wall, we were greeted by a slightly disappointed Roger, who could still hear the TV!

Roger was right. At 10:00 that morning, like clockwork, the TV came on, and the theme tune of *Homes Under the Hammer* was recognisable from next door. Having lived with the noise for more than twenty years, of course Roger was certain the noise was coming through the wall system he'd just invested in; but when he put his ear on the party wall he realised he could not hear the TV.

We walked to the exterior front window wall of the lounge. We didn't have to put an ear to the wall, but just lean slightly towards the window reveal. I remember Roger's surprised face in that moment when he realised the noise was transmitting down the flanking window wall perpendicular to the separating party wall.

We had already quoted for treating this indirect noise path. Once he'd heard it for himself, it was an easy decision – instead of plastering that day, we treated the window wall with the flanking wall system. We returned the next day and were greeted by Roger, this time with big smiles. He didn't hear his neighbours anymore! This was due to the combination of the flanking wall system with the thick multi-layered floating frame

STEP 3: TREAT THE INDIRECT NOISE PATHS

system installed on the direct noise path, which made it possible to reduce the noise significantly.

Treating the direct noise path of the alcoves alone will rarely give major reduction in noise. To ensure significant reduction, a floating frame needs to be installed and the flanking area treated too.

Next time you hear noise from neighbours, try putting your ear against a window wall. Do this for the different sections along the window wall and also for any wall joining a separating party wall, like a wall separating front and back bedrooms, for example. For those of you with road traffic noise, do this for the window wall but also for any wall at right angles to the direct noise path.

Those of you in flats and apartments with unwanted noise from above or below need to check all your walls. Next time you hear a constant noise from flats above or below, put your ear against the wall with the finger in the other ear and listen. It may be better to ensure you have little or no noise coming up or down the walls before investing in any soundproofing to the ceiling or floor.

If noise is radiating from flanking walls, it needs to be treated in a way that stops it resonating – like putting your hand on a ringing cymbal. The right products must be used, or you will channel the noise into other parts of the house or flat.

RSJs (rolled steel joists)

Sound travels quickly along steel (see table in Chapter 5). RSJs can cause noise problems in the following areas:

- Above a bay window on a semi-detached property
- Kitchen extensions, when the RSJ rests on the party wall
- Loft conversions
- Converted apartment buildings such as old factories converted into flats
- Where they are supporting a removed chimney stack
- In floors over walkways between terraced properties
- Large bi-folding doors
- Any open-plan living areas

Where any renovation involves knocking down a load-bearing wall and opening up a room, an RSJ is normally required. The noise problem occurs because RSJs are usually boxed in, which does not address the issue of noise transmission. An RSJ usually has an H cross-section, creating hollow sections when boxed in. Ideally the natural resonance of the steel needs to be addressed before these hollow sections are insulated. If you have two RSJs bolted together, then insulate each before bolting them together. Then the RSJ(s) must be isolated from the finished surface to be plastered. We always use a specialist panel system, which has a zero fire rating, is airtight and is fixed to the RSJ in the traditional way. This must all be installed so as to meet fire regulations specific to the type of building.

Sadly, a boxed-in RSJ is a commonly missed flanking transmission and details like these result in many soundproofing attempts having to be redone entirely, at great cost to homeowners.

Flanking noise

Let's take some examples of flanking situations and see if you can relate them to your own property or project.

Separating floor between a flat

It's very common for developers and builders to finish the wall, then install the separating floor afterwards, because it's easier and cheaper. However, it's terrible for sound transmission.

When installing new separating floor joists between apartments using wooden or engineered joists, it's important to fix the new floor joists to the wall *before* installing drywall plasterboard or any thermal insulation system to the walls. Otherwise, you are creating a clear channel for noise to pass between floors. New-build properties, barn conversions and new home extensions using structurally insulated panels (SIPs) almost always have this flanking problem. If your property has SIPs, you can expect plenty of noise transmission across the entire property.

Internal wall meets a ceiling or roof

In commercial buildings, function rooms, sports halls and most of all traditional office blocks, a partition wall between an open-plan office and a meeting room may only go up to a suspended ceiling. Poke your head through those suspended ceiling tiles and you see a never-ending void across the entire office full of ducting and data cables.

This approach is also used for separating walls between bedrooms and their bathrooms in new-build flats and apartments. It's rare for a developer to install extra joists to allow separating internal stud walls to continue up to the roof or floor above. As a result, noise is transferred across the whole of the property, guests get no privacy and it's very common to hear toilet noises in kitchen areas, and noise between bedrooms. For domestic soundproofing projects, you need to make sure the new soundproofed wall system continues past the ceiling, as high as you can go, to avoid noise from the neighbours' ceiling flanking over the top of your new soundproof wall.

Where the joists run into a party wall, or you have a joist running parallel to the party wall, it may be impossible to run the soundproofed wall system all the way up. Materials can be cut and shaped around joists, wires, pipes and ducting, tucking them in to reduce the flanking noise.

Flanking noise is the hardest noise to barrier, requiring persistence and the fiddliest cutting and shaping of materials. The importance of this part of the project is often difficult to communicate, since the client may be adamant that the noise is coming from the separating party wall, and not understand why the ceiling or even the window wall should also be soundproofed. Why would you spend a lot of money on a window wall if, like Roger earlier, you are convinced the noise is coming through the party wall?

Getting to the source of the noise

Any soundproofing project needs to focus on the noise source and its flanking transmissions.

CASE STUDY

George and Jane could hear the neighbours' sliding wardrobe doors clonking and clanging on their runners, and even coat hangers being taken in and out. These wardrobes were immediately the other side of the party wall in the main bedroom. The noise was affecting their sleep.

The neighbours had also renovated their kitchen, installing an RSJ, plasterboard, downlighters in the ceilings, wooden floors and plenty of hard surfaces that also amplified the noise.

After dealing with the direct noise paths, for the indirect paths we used a sound-deadening 50mm panel system for the party wall below, to prevent the kitchen noises transmitting *up* to the bedroom. The bedroom floor joists entered the party wall, so the bedroom floor was equally important in preventing the noise and vibration from transmitting through the structure, vibrating the bed and waking George and Jane up.

A floating floor was installed in the bedroom, isolating it from the structure and all the walls in the room. The flanking window wall, the separating wall between the front and back bedroom and all ceilings were also treated.

The banging wardrobe doors became the focus of the soundproofing project. The key to success was treating the indirect noise paths in the correct order.

This meant the floating floor to the main bedroom was installed last, after the walls were treated with soundproofing. The floating floor sat above the bottom of the walls and well into the room, properly isolated from all the walls. This ensured George and Jane got the maximum integrity from the interlocking materials used in the floating floor, and prevented the wardrobe door noise from getting into their bedroom indirectly by going down through the neighbours' floor and up the other side of the soundproof wall.

Although our clients could not afford to treat all the flanking areas downstairs, they're now able to sleep and wake up when they want to, undisturbed. When asked what they could hear after the installation, our clients explained that most of the noises were gone, including the dreaded wardrobe noises and all the kitchen noises.

Indirect noise paths must be treated in the right order, otherwise some may still be present and you may not achieve the full benefit from the sound insulation materials used.

Flanking noise through floors

Where an internal wall meets a floor, one option is to continue the wall down to the subfloor. This is also commonly found with recent renovations using stud partitions placed on concrete floors that continue through to the next room. The worst scenario I've seen was a newly built, semi-detached housing development that shares a concrete slab with the neighbours. This means every homeowner can hear everything their neighbours do, especially impact noise: furniture scraping over the tiled floor, the vacuum cleaner hitting the skirting, light switches, plug sockets, doors slamming and roller blinds being pulled up and down. In this case, it was unfeasible to soundproof the property because it would take a complete renovation to adjust the floor heights and every floor surface of the house.

Continuing the new separating wall down to the subfloor is a basic soundproofing principle that prevents noise passing under a wall system. If you're unable to do this, separating the floor so that it's not continuous under the wall system is an alternative. In new-build properties, developers often install all the floors first, using chipboard, then install separating stud walls between bedrooms and bathrooms. This causes massive noise transfer between rooms.

Flanking noise through the loft

One of the best examples of flanking noise is in semi-detached bungalows when the party wall separating the two bungalows does not continue up to the roof: they share one big loft. TV noise can transmit up through the ceilings of the bungalows and down into the living areas below (or next door). The neighbours have no privacy and find themselves competing on TV volume.

Lofts present many flanking issues, which is why a cheap loft conversion can amplify and channel noise through the rest of the property.

In most cases, it's not practical (even with all the persistence in the world!) to continue a soundproofed wall system up through the bedroom ceiling and into a loft. You'll have structural joists and purlins in the way, or even a chimney stack. To reduce this kind of flanking noise, you have to try to continue as many layers up past the ceiling as you can. Then, in the loft above the ceiling, either continue the wall system up or insulate the entire loft.

Remember, thermal loft insulation does nothing for sound attenuation. Acoustic mineral wool slabs cut and positioned between the joists, filling every void, would be better. You can, if you choose, put the thermal insulation back over the sound insulation, but don't block up the eaves or install acoustic mineral wool insulation between the rafters of the roof unless you've had a specialist's consultation and advice, as you could cause condensation.

Chimney stacks

As chimney stacks must be able to withstand heat, they are commonly made with a double layer of fire bricks. These bricks tend to be denser and thicker than standard bricks and it is widely believed that these two layers are adequate to stop noise transfer from a neighbouring house. We have found this not to be the case in hundreds of homes and strongly recommend the entire chimney stack is soundproofed. Simply put your ear on the chimney stack (as I suggested you do for window walls). As a flanking area, the chimney stack radiates structure-borne noises and therefore needs to be 'sound deadened' using high-quality acoustic materials (which should carry a zero fire rating).

If the neighbours remove a chimney stack you share with them (which normally takes one day, so can happen without you knowing), unwanted noise will increase.

Summary

If flanking transmission (AKA indirect noise paths) is not properly soundproofed, putting a state-of-the-art soundproofing barrier on a party wall, floor or ceiling alone may achieve very little. A direct barrier will change a room's acoustics when standing right in front of it, and may channel the noise into another room, but it won't address indirect noise paths. You will not get the full benefit of soundproofing rooms upstairs until you've also treated the rooms downstairs.

Does your window wall need treating? Or the wall separating the front and back bedrooms? Put your ear on the wall, and if you can hear any structure-borne noise coming down that flanking area you will get an idea how noise is coming through indirectly.

Sound travels well along steel. Make sure any RSJs are sound-deadened and insulated.

I strongly recommend that a chimney stack on any separating party wall is soundproofed. Simply put your ear on it (as above), and if you can hear unwanted noise on the flanking area, then you need to soundproof this area.

Make sure you tuck any soundproofing up above the ceiling area if possible and tuck any system down below the floor level if possible.

A floating floor must go up to but not touch the walls or the skirting. Otherwise, you are creating a clear channel for noise to pass between floors.

Flanking noise is not always considered by architects and developers, and building regulations focus on direct noise paths – either the separating wall, or the separating floor in apartments. Go that extra mile and soundproof that ceiling and flanking window wall. Insulate that RSJ in a way that meets fire regulations. Make sure you insulate the floors and ceilings and flanking walls to get the best from your investment. For many years now, I have not soundproofed a home without insulating the ceiling and floors. There is a reason for this. Use this book as a guide to tick off all the flanking indirect noise paths and make sure every area has been addressed to get the best results.

7
Step 4: Treat The Fixtures And Fittings

Treating the fixtures and fittings is a key area of any soundproofing project that is always overlooked. Every week I speak with at least three people that have had a soundproofing attempt of some kind. With all of them, the fixtures and fittings have been ignored and the soundproofing has been compromised.

Plug sockets

Take plug sockets, for example. To fit a flush double socket into a party wall, we see builders cut a hole the size of the socket box into the soundproofing layers that they have applied, thinking, 'It's such a small section, such a small bit of the wall... it won't make a difference'.

In fact cutting a hole the size of a double socket in your new soundproof wall system will make a massive difference to the insulation: we have found through acoustic testing that a hole the size of a double socket on a separating party wall can mean failing the required acoustic test (42dB DnT,w + Ctr – see Chapter 1 for these details). When we corrected the builder's mistake and installed the socket correctly, we achieved 47dB DnT,w + Ctr (a pass, though a narrow one; the higher the figure the better).

On a separating party wall, compromised double plug sockets are most commonly found behind the TV, or in an alcove where you position your chair next to the separating party wall so you can sit facing the TV. In kitchens, plug sockets are positioned above the work surface, right where you make your cup of tea. In the bedroom, these compromised plug sockets are usually each side of the bed. These are all areas where you do not want any leakage of sound. The sockets should not be the weakest points on the direct noise paths; try to make them one of the strongest parts of the wall system.

The materials used around plug sockets need to be the same if not better than the acoustic materials used in the wall system, so it is very unlikely an electrical wholesaler will supply anything strong enough for the purpose.

Downlighters and recessed ceiling lights

Downlighters and recessed ceiling lights are installed by cutting a hole in the ceiling board and inserting the lighting unit. The units do not provide any acoustic insulation, so these holes are a huge problem when you have downlighters positioned in the ceiling of a lounge or dining room that has a conflicting separating wall. Noise will simply come over the wall, through the ceiling void and out through the holes cut into the ceiling.

If you have noise from neighbours and you have only treated the direct noise path (the conflicting party wall), often it is difficult to pinpoint exactly where the noise you are now experiencing is coming from. Downlighter holes in the ceiling can make noise from neighbours sound as though it is coming from all around you.

Either the recessed downlighters need to be exchanged for a single light pendant, or special acoustic boxes need to be installed over the top of the holes in the ceiling. This work has to be done from the floor above, pulling up the floor in the room above, locating the downlighter holes and fitting acoustic downlighter boxes to each of the light units that penetrate the ceiling. Using acoustic insulation can cause many lighting units to overheat and the bulbs to keep blowing, so it is important a box is used. Insulation alone is not enough to prevent noise transmitting through the downlighter holes. Acoustic downlighter hoods sold with a fire rating are better than nothing, but the best results come from using bespoke acoustic downlighter boxes that are acoustically sealed to the ceiling board and installed from above (see 3-part soundproofing documentary series on www.quietco. uk/videos before and after soundproofing a semi-detached house).

A lot of newly built properties offer downlighters as an option in the specification. These are no doubt installed in uninsulated ceiling voids with chipboard floors above, making it very difficult to get to every downlighter hole without removing *en suite* bathroom furniture and floors.

The situation is even worse for flats and apartments, where the only alternative may be a new lighting design using wall lights and single lighting pendants hung from the ceiling of the room affected by noise. If the apartment has a concrete ceiling, a suspended ceiling must be installed beneath the ceiling after soundproofing, in order to have recessed downlights. If this is not possible due to height restrictions, then new lighting design is the only option.

Fire designs

As you have gathered after reading this far, sound-proofing the entire chimney stack (Step 3) and the fire back behind your fire (Step 2) are very important parts of the four-step method. The fixtures and fittings are the actual fire feature, and this must be installed in a way that does not compromise the soundproofing and complies with Document J of the Building Regulations. As I mentioned in Chapter 5, the key bit to note is that you cannot have any combustible material within 200mm of the flue for the fire. There is a huge choice of fires and surrounds and I have not come across any design that cannot be installed with soundproofing. A specialist stove or gas-registered engineer would be able to advise on the other installation regulations specific to your chosen fire design. Electric fires do not have to meet Document J and so these can be installed with a wider range of soundproofing materials used on the fire back and chimney stack. Thus, choosing an electric fire can mean a greater reduction in noise from neighbours, for example.

Coving

Coving is used for decorative purposes to cover the join between walls and ceilings. It is often used to cover up something like a wavy ceiling or cracks where the ceiling and wall join. Standard, off-the-shelf coving will usually have void behind it, which will channel noise around a room. Like any wind instrument the noise is louder at the end opposite the source. This modern coving is connected to the top of the wall and to the uninsulated ceiling void so any noise can be transmitted from the walls, around the room and straight into the ceiling void and any upstairs rooms. Coving should be taken off and not refitted unless it is absolutely necessary; if so, it should be fitted in a way that does not compromise a resilient ceiling or wall system.

CASE STUDY

Jackie lives in an end-of-terrace property. All the rooms in her house had coving.

We soundproofed one room for her, and she asked us back a few months later to soundproof the front room, so she could watch her TV without having to hear next door's TV on delay. It worked: she simply did not hear the neighbours, TV, conversation, music and, best of all, she was no longer disturbed by the neighbours' front door slamming.

Jackie understood the reasons not to refit the coving after soundproofing, but did not want the rooms to look out of keeping with the others. She decided to ask a decorator to put the coving back up in the two rooms she had soundproofed. I received a phone call and she told me, 'Jim, I hear the TV and muffled conversations.' Jackie had made the tops of her new soundproofed wall negative: it could no longer vibrate where the coving was fixed and therefore could not convert sound energy into heat. The new coving she installed was also channelling noise around the room from other parts of the house. She got her husband to remove the coving and we plastered the damaged areas for her... and, once more, she no longer hears the neighbours.

Floors

Hardwood floors in homes are very practical and offer great thermal qualities as well as clean, dust-free environments for allergy sufferers. However, they do not perform very well acoustically and are often banned in many flat tenancy agreements. They amplify impact noise such as footsteps and furniture moving and, compared to other floor finishes such as carpet, hardwood floors do not absorb much airborne noise such as conversation, music and TV.

A couple in London faced a legal bill of £1 million after losing their battle over a noisy wooden floor.[28] The couple lost a court fight with their downstairs neighbour, after she complained of 'intolerable' disruption from their apartment. The neighbour downstairs said the couple's three children treated the home 'like a playground' and sounds of them passing over the wooden floor, as well as noises of dishes being washed, kept her awake at night. I believe the noise may have been caused by the construction, or the lack of sound insulation, or the way the fixtures and fittings were installed, and in particular the way the wooden floor was fitted. The neighbour below also said that the peace and tranquillity of her £2.7 million home had been ruined by noise from a boiler, a fridge,

---

28 Jameson, S, 'Noisy neighbours must pay banker £100,000 in compensation because their floors had no carpet, judge rules', *The Telegraph*, 5 March 2018, www.telegraph.co.uk/news/2018/03/05/noisy-neighbours-must-pay-banker-100000-compensation-floors, accessed 16 October 2020

taps and a fireplace – all important fixtures and fittings that may not have been installed in a way that reduced noise transfer.

As well as causing annoying structure-borne noise to neighbours, hardwood floors reverberate themselves and can reduce the sound insulation value of a room. Adding carpet to a room with a hard floor can improve the acoustics within it, and through absorption can reduce impact and airborne noise transmitted to the floor below. Installing carpet and underlay on bare floor boards will give a noticeable reduction in noise from next door. We would always recommend treating a floor with carpet and underlay if practically possible.

All floors covered with (or made of) wood, laminate, marble, ceramic, concrete or stone must be isolated from the structure. A hardwood floor needs to be isolated from the walls to ensure vibration from the floor does not transmit to them; a 5–10mm isolation gap is sufficient (the skirting will cover this join). Impact noises like furniture moving can transmit through a hardwood floor and straight into the structure of the building if it is not isolated.

Finished floors in flats and apartments must be isolated from all the walls in that room and isolated from any joists, too.

If your floor is made of wooden joists covered with chipboard or floor boards, then you must insulate between the floor joists, cut open that chipboard floor, or pull up those floor boards and insulate with acoustic mineral wool. This is an absolute must to reduce the noise. A floating floor has segments of acoustic board connected together with resilient layers, and is normally about 40mm thick. Floating floors are not glued or screwed to the joists and isolated from all the walls in the room. The finished floor ideally needs to be installed over the floating floor. This means noise will find it very difficult to transmit from a hardwood finished floor to the joists. It can reduce impact noise significantly: we have found that clients perceive footfall noise reduced by over 80% where a premium floating floor has been installed.

For apartments with concrete construction, the finished floor ideally needs to be isolated with a resilient layer. If a hardwood is to be laid over it, ensure any heavy-duty underlay is durable enough. Too much movement in the underlay could mean joins in the hardwood pinching your feet if you walk over them in bare feet. A 'cradle and batten' floating floor system is normally the preferred acoustic treatment for a concrete floor. It offers the option of installing acoustic mineral wool to the separating floor. They come in a range of heights but normally will raise the floor by at least 80mm. Remember, though, with a concrete construction, 50% of the noise will come down the flanking areas, most likely the walls (see Chapters 5 and 6).

Radiators

Heavy radiators need to be considered during the design phase: is it better to move the radiator to a new location? Perhaps replace it with a space-saving vertical radiator? Whatever your choice, if radiators are connected to a party wall to be treated, they need to be taken off and reinstalled in a way that does not compromise the soundproofing. Of course, all the holes where pipes and screws and brackets penetrate the wall need to be sealed but, more importantly, how the radiator is connected to the wall and/or floor could make the resilient wall system negative in this area – like Jackie's coving. The radiator therefore needs to have its own resilient channels installed in the wall system. The radiator is then fixed to these resilient channels after painting and decorating. It is easily done but must be taken into consideration in the design phase.

TV mounted on the wall

Fixing TVs to a party wall is most commonly achieved using a TV bracket suitable for the specific TV. These brackets are screwed directly to the plastered wall and can cause huge discomfort for any neighbour the other side of the party wall, who will most likely hear the TV noise in every room. They will struggle to get away from the noise, mainly because the TV noise is being transmitted into the structure of the building. Like a subwoofer or stereo speakers positioned on a suspended floor void, low frequencies will transmit through the structure giving the neighbour no escape from the noise, unless they soundproof all the indirect noise paths (see Chapter 6).

Extra resilient layers need to be installed within the new soundproofed wall and behind the TV bracket too. The fixings that hold the TV bracket also need to be decoupled from the bracket using anti-vibration pads to reduce the vibration transmitted from the TV to the wall surface.

Doors

When your neighbour's door slams, you hear the impact of the door leaf hitting the door frame, and may also hear the door lock hitting a strike plate on the door frame. Both these impact noises transmit into the structure of the building and all the way to your property.

The impact noise of doors slamming is a difficult noise to stop in any situation; with apartments and flats it can be one of the most annoying noises that invades your space. The sound of people coming and going can disturb an evening, the enjoyment of a good film or even affect your sleep.

One way of reducing this noise is to reduce structure-borne noise by installing sound-deadening matting to any surface connected to the structure the door is hung from. In most cases the noise of a door slamming can be considerably reduced but will still come through the best soundproofing.

The best way is through isolation. With many homes, we find that disconnecting the floor joists entering the party wall on a property can reduce the noise of the neighbour's front door (as well as children running around and other impact noise). The noise is transmitting through all parts of the structure and they all need addressing by sound deadening and isolation in order to get any sort of reduction.

Treating the banging door noise at source

Acoustic seals installed to the door frame will give any door a softer close. The use of magnetic door latches will eliminate the noise of the latch hitting the strike plate and any mechanical noise of a handle opening the door. However, magnetic latches will present problems for final exit doors, which require a certain level of security while still being easy to open.

For the considerate noise maker with a noisy hobby, hanging a single-leaf acoustic door to a home office or studio usually achieves a value of 44dB if we use a frame that has three sets of acoustic seals and a 64mm-thick door leaf. If a greater reduction in noise is required, then a two-door system can be installed: two single-leaf doors, with one opening out towards you as you enter the room and the second opening into the soundproofed room and a 200mm gap between the two leaves. With our acoustic double-doors system, we normally achieve a 60dB reduction in noise. These door sets are very effective at reducing acoustic drums, and heavy bass music. I have designed and built hundreds of home studios, in garage conversions, garden studios or a room in the house; most require a double-door system.

The increased popularity of working from home has increased the demand for multi-purpose soundproofed rooms in the house: one room soundproofed to a very high standard, reducing noise from the rest of the house, from neighbours, environmental noise, where you can have private conversations and make as much noise as you like whenever you like. A really thick acoustic door should allow you to shut off noise from the rest of the house. During lockdowns, we experience an increase in enquiries for this type of room within homes.

Kitchens and bathrooms

Other less obvious finishing issues arise when fitting new wardrobes, new kitchens or new bathrooms to the new soundproofing investment. Ensure the connections made do not compromise the soundproofing. Making a fixing to the new isolated and resilient soundproofing system and then attaching it to a fixed surface can strip a soundproofed ceiling, wall or floor of its resilience. For example, installing a new kitchen along a soundproofed wall: the work surface and wall units are usually fixed to the walls and the base units to the floor. The wall system will no longer be able to vibrate and will therefore not be able to covert sound energy into heat. For kitchens, use a system that is not dependent on resilience; for instance, a wall system that deadens resonance, has mass that can take fixings and still offers the required reduction in noise.

CASE STUDY

Elizabeth runs a massage therapy business from home with her own treatment room in a beautiful self-contained flat. The treatment room is below Elizabeth's kitchen and lounge. Her young children and family would walk about upstairs and the impact noise would disturb the patients having therapy below. The business started getting really popular and the problem got worse for the family, who were asked not to use the kitchen and living area during therapy sessions.

We removed the kitchen and reinstalled it on top of a floating floor isolated from the floor joists and all the walls in the room, which massively reduced the impact noises of pots and pans in the kitchen, the children jumping around and general living noise being heard downstairs in the treatment room.

The reinstallation of the kitchen after soundproofing is a great example of installing fixtures and fittings in a way that does not compromise soundproofing. There is a lot of movement in a floating floor and we did not want Elizabeth's new work surface to move up and down as someone walks across her new floor. The main floating floor runs up to the legs of the base units, which sit on their own, isolated floating floor, and the unit kick plates cover the join.

Elizabeth simply does not hear noise from upstairs during therapy sessions, and no longer has to apologise to her clients during treatment.

Sound absorption

Often confused with soundproofing, sound absorption is achieved after the soundproofing by fitting material like high-performance foam to absorb sound energy that is then lost to heat.

Soundproofing is the *reduction* of sound energy as sound passes through a barrier, using sound insulation to reduce the noise entering or leaving your home. Sound absorption softens echoes and reverberation in a room after soundproofing.

Sports halls, church or village halls, and community centres are not the only places that suffer from echo and reverberation problems. Large kitchens with many hard surfaces can create acoustically uncomfortable environments. Echo in a newly soundproofed lounge without soft furnishing can cause headaches, fatigue and make the room uncomfortable to live and entertain in.

CASE STUDY

Judith and Brad renovated a beautiful old church, building a six-bedroom extension off the back. The problem was echoes in the open space and so many hard surfaces.

The solution was installing large printed tapestries on specialist fabric, which fit well with the renovation. The specialist fabric is full of tiny holes that lets the sound pass through to high-performance glass fibre that absorbs the sound energy. Judith and Brad noticed the difference in the acoustics immediately. They decided to have other pictures printed and mounted the same way and scattered them elegantly around ornaments. This massively reduced the echo and reverberation in their home and now the open-plan living area is a pleasant, acoustically comfortable environment.

If you are having a large room soundproofed, you may end up with more highly reflective surfaces in the room. Consider using soft furnishings that will absorb some of the noise.

I often get asked, 'Can I put egg boxes on my wall to reduce the noise from next door?' These do nothing for sound reduction but will change the acoustics within the room, so the noises you hear sound different. The difference may convince you that the noise has reduced, but what you have done is change the way the sound energy bounces around in that room. The egg boxes will act as a sound absorber, but a good sound absorption material is not necessarily a good material for soundproofing. A good soundproofing material is heavy and airtight and not a good sound absorber.

The reduction in reverberation by adding sound absorption material depends on a number of factors:

1. **The size of the room:** In particular the height, because reflections off the floor and then the ceiling are the first reflections we hear; they are normally the surfaces we are closest to and therefore a good area to treat with absorptive materials.

2. **The sound absorption qualities of the existing contents:** Hard surfaces, like walls, hard floors, tiles and marble, are all highly reflective and should have some treatment. Carpet is the obvious choice for the floor, immediately reducing the echo in most homes, but sometimes it is not practical (eg for an open-plan kitchen area).

3. **The location of the noise source:** This can be hard to identify when the sound is travelling to the human ear from lots of different directions within a room.

4. **The positioning of the new sound absorption material:** The more absorption you install on the hard surfaces of a room the shorter the reverberation time becomes; sound reflection reduces, which improves the acoustics in the room.

5. **Types of absorptive material:** If you have canvas pictures, you could install glass fibre or high-performance foam in the frame behind them to reduce echo and reverberation in large rooms that have a lot of hard surfaces. Acoustic mineral wool is a very good sound absorber, but not very aesthetically pleasing and the fibres can irritate. Sofas, rugs, curtains, drapes, beds, soft chairs and cushions are all good sound absorbers.

Acoustically live

This means the room has a lot of echo and sounds a bit like an old sports hall or church. This is caused by the number of hard surfaces in the room. Sound reflecting off hard surfaces can make it difficult to distinguish words in conversation.

Acoustically dead

An acoustically dead room is like an 'anechoic chamber', those rooms that have giant cones of foam rubber all over every surface. It is very noticeable when you stand and talk in an acoustically dead room; it feels strange to your ear and difficult to balance. The first thing you notice in an anechoic chamber is the blood flow in your ears. These rooms are often used for testing; for example, car manufacturers test how much noise the fluid in their electric car batteries will make, slushing around while driving. Manufacturers of luxury cars also use these rooms to test how much noise is made when closing the glove box.

Home office or home studio

For your home studio, it is sensible to have some hard surfaces and get the balance right: neither too 'acoustically live' nor too 'acoustically dead'. For our clients, we normally treat the entire studio ceiling area with a high-performance sound absorption material and then leave the client to add more absorption and diffusion products if they require. If you take video calls or conference calls at home, getting the acoustics right makes a massive difference. Strategically positioned pieces of specialist sound absorption where you take phone calls will make it easier for your speech to be heard by the other parties on the call.

Building Bulletin 93 (BB93)[29] explains minimum performance standards for the acoustics of school buildings, including legislation on the control of noise and reverberated sound in schools and other public buildings. Sound absorption is important to get right and should be considered with every soundproofing project, to make the home acoustically comfortable.

Summary

Fixtures and fittings seem so obvious, especially socket boxes cut into the direct noise path or down-lighters cut into key flanking ceiling areas, yet they are always missed. Some fixtures and fittings are not so obvious: coving or a built-in wardrobe, for example. A wardrobe fitted with an uninsulated void behind it can channel the noise between floors and also amplify noise in the same way 'dot and dab' plasterboard does.

If you have a chimney stack on a party wall, most electric fire designs do not have to meet Document J of the Building Regulations and so a wider choice of sound insulations can be used on the fire back and chimney stack; so, choosing an electric fire can mean a greater reduction in noise from neighbours.

---

29 See www.gov.uk/government/publications/bb93-acoustic-design-of-schools-performance-standards, accessed 17 December 2020

Take the time to design and plan your room before you start pulling up that floor and removing that ceiling. Do the same as you would renovating a kitchen or bathroom; purchase all the bits first before you start the installation.

Hardwood floors do not perform very well acoustically and are often banned in many flat tenancy agreements. They amplify impact noise such as footsteps and furniture moving and offer less absorption of airborne noise such as conversation, music and TV than, say, a carpet. Adding carpet to a room with a hard floor can improve the acoustics within a room.

All floors with a hard finish (wood, laminate, marble, ceramic, concrete, stone) must be isolated from the structure. Impact noises like furniture moving can transmit through the floor and straight into the structure of the building otherwise. In flats and apartments finished floors must be isolated from all the walls and also isolated from the joists. The finished floor ideally needs to be installed over a floating floor.

Ensure fittings connected to new soundproofing do not compromise it.

Sound absorption should be considered after soundproofing larger rooms because the increased number of reflective surfaces will increase the echo within the room.

Consider acoustic comfort, especially if you take video calls or conference calls at home; it will make you feel less fatigued and make conversation easier.

8
Are You Ready To Benefit From Soundproofing?

I'm going to look in detail at a case study that pulls together many of the points I've made in different parts of this book.

CASE STUDY

Jane and Rob lived in a 1950s three-bedroom semi-detached property, the type of house that has a chimney stack in each room along the separating party wall and when you remove the wallpaper it all goes back to brickwork and has to be completely replastered. They extended the back of the house, which involved installing a large RSJ across the back of the house that sat on the separating party wall. The extension was fitted with recessed downlighters, underfloor heating, engineered oak flooring, bi-folding doors and a TV on the wall.

After the renovation, they could hear their neighbours' dogs barking, music every evening, they could make out words in conversations and hear doors banging, plug sockets, light switches and the neighbours' TV.

Unfortunately, they had not considered soundproofing before building the extension and the wrong materials were used. They did some research into soundproofing and realised they might not be able to afford all the soundproofing work that could be needed. So they got a friend to help them install a wall system on the party wall alcoves and the chimney stack, add sound-resistant plasterboard to the new extension walls and insulate the ceiling and the suspended floor void.

After painting and decorating their new extension for the second time, they were absolutely gutted to hear conversation from the neighbours; they noticed a slight reduction in noise from the party wall alcove, but could still make out words. In the bedroom above, noise had also been made worse. The strangest thing, to Jane and Rob, was that the wall of their new extension does not share a party wall with the neighbours' house: it extends out into the garden from the back of their house.

Even worse, the neighbours started to complain about the noise their kids were making, noise from the TV on the wall and from the bi-folding doors banging.

Jane and Rob's new extension could not be used because they heard their neighbours and the neighbours were getting annoyed at hearing them. Rob started to get concerned about the lack of privacy.

A year later Jane and Rob contacted us. They paid for a soundproofing consultation visit, giving us access to the neighbours' property, so we could test and show Jane and Rob where the noise was actually coming from. It was actually louder on the indirect noise paths: along the RSJ, the new extension's outside wall and the separating wall between the lounge and kitchen. As this was the third time the extension would have been renovated, we had the current sound insulation tested by an independent, UKAS-accredited acoustic testing company.

The sound insulation of the *party* wall measured at a surprising 53dB DnT,w + Ctr, 8dB higher than the required standard for properties newly built today (the higher the figure, the better the insulation – see Chapter 1 and Appendix). However, we could still hear conversation through the *extension* wall.

How is this possible?

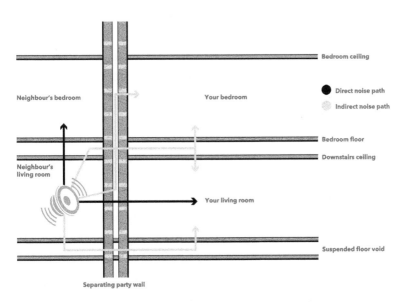

Noise in a downstairs room. Adapted from Rindel, JH, Sound Insulation in Buildings *(CRC Press, 2018)*

THE NOISE FREE HOME

The attempt to soundproof the party wall by Jane, Rob and their friend had done what the wall system said it would for the direct noise paths, but unfortunately the way it was installed using traditional building techniques channelled the noise into indirect noise paths such as the ceiling, floor, extension wall and separating lounge wall (flanking areas). Certain frequencies of noise were also still able to come though the wall system. The sound-resistant plasterboard added to the extension wall had been fixed using the 'dot and dab' technique. Jane and Rob had hoped that such heavy board would make things better, but in fact they had made the noise from the neighbours worse.

At certain frequencies there can be weaknesses and because the UKAS sound insulation test is an average over a wide range of frequencies you can get reasonable test results, yet still hear noise. The testing does not take into consideration real-life 'clonks and clangs' from neighbours or noise transmitting across the house.

There was a lot wrong with Jane and Rob's second renovation: they did not isolate the finished walls from the aerated blocks used for the new extension walls, they fixed the plasterboard using 'dot and dab', they used rigid insulation in their sloping roof, and the soundproof system was not installed by a specialist installer. The wall system they applied to the party wall gave a fair amount of sound insulation when you stand directly in front of this wall, but channelled noise into other areas.

But some of the problems went back to the initial build. Boxing round the RSJ amplified certain frequencies of noise. Voids in the structure (eg the ceiling void) let in higher-frequency noise from their neighbours (small sound wave). The recessed downlighters allowed noise to pass indirectly through the ceiling (flanking noise). Sockets, skirting, the engineered wooden floor and the TV bracket had all been installed in a way that created holes in the structure and compromised any sound isolation. The way the fixtures and fittings were installed amplified noises like furniture moving, doors banging, plug sockets and light switches as well as low-frequency bass noise.

This is why, although Jane and Rob's sound insulation value far exceeded government standards, they could follow their neighbours' conversations.

Applying the four-step soundproofing method to Jane and Rob's home

After removing all the soundproofing from the party wall and taking it all back to the brickwork, we treated the:

- **Main cause of the problem:** 'Dot and dab' plasterboard fixing and holes where the joists entered the party wall

- **Direct noise paths:** Party wall alcoves and fire back

- **Indirect noise paths:** The uninsulated RSJ running into the party wall, the extension wall and ceiling, the kitchen diner ceiling, the suspended floor, the chimney stack and the separating wall between lounge and kitchen diner

- **Fixtures and fittings:** The downlighters, the wooden floor, the TV mounting, plug sockets, skirting, and shelving and some sound absorbing pictures

Jane and Rob do not hear their neighbours anymore. The neighbours do not hear the children and both households have their privacy.

After installing the soundproofing, we got the extension tested again by an independent company that carried out a UKAS-accredited test. The result measured at 62dB DnT,w + Ctr, 9dB better than the result before we carried out our installation. This 9dB difference means that almost half the noise that was getting through before is now no longer getting through. This is comparable to the soundproofing of a detached house, and not just on the party wall but for the main cause, flanking indirect noise paths and noise transmitted via the fixtures and fittings.

What was involved?

To carry out the installation, we moved Jane and Rob's furniture into storage; soundproofed their extension; included in all plastering, electrical work, carpentry, painting and decorating; then we moved them back in and put all their pictures and fittings back for them too. The cost of the final renovation was thousands of pounds, but for Jane and Rob it worked out considerably less than moving to a detached house and was less disruptive, too.

Some home truths you will not find online!

There is no such thing as 100% soundproofing despite what you might hear. If your home is attached to another property, you will at some point hear your neighbours. With the most effective soundproofing on the market today, you will still hear your neighbours if they have a party or play bass-heavy music at an unsociable level when you are sitting in silence next door.

If your neighbours are shouting and screaming at the tops of their voices and you are sitting in silence, you will hear these noises; but good sound insulation, properly installed, will considerably reduce them. (It also depends on what sound insulation materials have been used, the type of property and the level of noise.)

If your neighbours are knocking or tapping on the separating wall, ceiling or floor, impact noise will still come through the majority of sound insulation systems.

You will not get the peace and quiet of a detached house by treating only the alcoves either side of a chimney stack.

You are unlikely to feel the full benefits of soundproofing an upstairs wall until you have treated the downstairs wall.

When treating a separating floor from above, you are usually only installing half a system: the other half needs to be fixed to the ceiling below.

To get a significant noise reduction, you need to follow all four steps I have described: doing two or three steps is not good enough. You will be disappointed with the results, and question the disruption and loss of space.

Although every house is different, soundproofing your home properly can be life-changing and reduce an amazing amount of noise. Unwanted noise cannot be completely eradicated but by following the four-step soundproofing method it can be reduced to an acceptable level, a level at which you get your life back. Following all four steps will make sure you obtain the full benefit of the materials you invest in.

After soundproofing, most of the time the noise you hear will be from outside, through the window.

CASE STUDY

Sally lived in a 1930s semi-detached property and before soundproofing was woken up every night by doors slamming next door. She also suffered from neighbours' music, TV, conversation and children running around. We tested Sally's property after soundproofing and it achieved a sound insulation value of 63dB DnT,w + Ctr. After the Christmas holiday period, we called Sally and asked what she had heard: she thought she had once heard a door slam but was not sure if it was outside or not.

Your own privacy

After soundproofing, whether you're normally noisy or normally quiet, you won't have to worry about your noise disturbing others, you will stop noticing your neighbours and start to feel relaxed and comfortable in your home. Perhaps you will start to do things you did not do before: have friends round, say. You will notice that you are not getting woken up by noise, sleep better and no longer go to bed feeling stressed, annoyed or frustrated at the neighbours.

Our clients, Jane and Rob, had a great relationship with their neighbours and had the privilege of being allowed to test the party wall between their houses before and after soundproofing. Their neighbours understood the benefit the soundproofing would have for both households. The most rewarding part for Jane was when she realised that she could sing and play with the children and that their noise was just not audible in their neighbour's house. For Rob, the most impressive part was the shouting tests: everyone had a go at shouting through the party wall to see if it could be heard next door. It simply was not audible unless you had your ear right on the party wall.

Your health improves

Every time you hear unwanted noise your blood pressure increases and feelings of stress and annoyance flow. Once you no longer hear them, you will not have those feelings anymore.

You will enjoy your house more and feel you can express a range of emotions, maybe listen to music, play your favourite song nice and loud, perhaps have a dance and sing along. This leads to positive self-esteem, and that leads to better relationships with those around you.

You feel settled in your home

After soundproofing, some clients spend more time in their home; they enjoy evenings at home rather than trying to escape unwanted noise. Some clients have more guests round to stay, and they entertain more in their home. Rather than longing for that 'other' house or flat where noise problems won't exist, after sound-proofing they never think of moving again. You might start to feel excited when you arrive home, because during the design phase of the soundproofing you can think about building in things you have always wanted: perfectly positioned doors, walls, shelving, cupboards, storage, lighting and finishings, say.

You feel that you can do what you want without both-ering your neighbours. If you invest in a soundproofed office, studio or multi-purpose room you will be able to make as much noise there as you want, whenever you want, giving you freedom to do what you love in the home you love.

What is it like not hearing road noise anymore?

Imagine being able to look out your windows and see the cars go past and not hear the noise from the traffic; not hearing the hum of the motorway or dual carriageway, the rattle of the railway. Do you think, 'That's how loud life is'? Why? Architects and builders have not used the latest materials, they have not been taught soundproofing techniques, and they have not stayed up to date. There is another way. If you live near a busy road or railway, you can soundproof your home. Your overall health will improve; your children's concentration will improve.

CASE STUDY

Zoe has a beautiful house right on a busy road near Birmingham. A particular annoyance was lorries driving over drain covers. She was promised a massive reduction in noise from installing double glazing, but unfortunately the installers did not fit the windows tightly enough (see Chapter 5) and standard double-glazed windows are not adequate for high levels of road noise.

Our in-depth consultation found the noise was coming from the following areas of the property:

1. Roof rafters: not insulated **(main cause)**.
2. Poor glazing: not double-glazed, laminated acoustic glass **(direct noise path)**.
3. Plasterboard amplifying the noise on the window reveals, the conflicting road wall and flanking walls; and the floors in the bedroom not insulated **(indirect noise paths)**.
4. Poor fitting of beading around the windows, downlighters in the ceilings and the window sill; coving channelling noise around the bedroom; skirting and finished flooring not isolated **(fixtures and fittings)**.

Typically, the recently renovated roof used rigid PIR boards, good for thermal insulation but giving no respite from road traffic noise (see Chapter 4).

We addressed all these areas, and fitted new secondary glazing to Zoe's daughter's room. Zoe no longer hears road noise in her daughter's bedroom. The main bedroom is also much better, though Zoe does have to close the door to the main bathroom at night, as she could not afford to soundproof the bathroom as well. In a courtesy phone call two months after installation Zoe told us she had not heard the drain cover once.

Why is the expected level of noise not on the survey?

Unwanted noise in your home affects your health and broken sleep will raise blood pressure, having a detrimental impact on all aspects of your life and the relationships with people around you.

It does not have to be like this. Moving to a detached house is an option to escape noise from neighbours but may bring about other noise problems. You may still not be able to play your instrument when you want and noise from that road or a neighbour's barking dog will still affect you, for example. Many of our calls and enquiries come from detached houses built from 1990 to date or recently renovated detached homes. The most popular noise problem is noise transfer between rooms, such as toilet noises or hearing TV and conversation throughout the house.

After soundproofing my own home and hundreds of other buildings with a wide variety of noise problems, I believe the unwanted noise in your home is probably caused by the construction of the building. Houses are not built to address modern noise and noisy gadgets. I believe we can change the way we build our houses reducing the noise in homes.

My team and I are hoping to change British building standards for sound insulation. Through raising awareness of the problems of noise and the effects it can have on people's lives, we will press for a new requirement. Every house should have a noise rating survey when it's sold, so potential purchasers know whether they will be able to enjoy peace and quiet in their home, and pressure will be put on architects and builders to use methods that do not create noise problems.

When you buy a new house, regardless of its age, you usually request a structural survey, an energy performance certificate, an electrical certificate, a boiler service certificate; and any major alterations will have a building certificate. You will even receive a window insulation certificate. A home survey will outline any damp or dry rot problems and even describe the level of loft insulation.

Level of noise from the road can easily be measured. Noise from neighbours could be measured too, over a particular period: say a week. Noise from any new alteration can be outlined on a survey. Why is noise not considered on the house survey? People are just not talking about it. Help get the conversation going on Twitter, with the hashtag: #whynosoundsurvey.

Although a surveyor will not look behind furniture, or lift up floor boards, or drill any holes, they will tell you about any unstable walls, subsidence, the thickness of the loft insulation and the quality of the windows. Understandably, a survey can be very reassuring to a new buyer, but a lot of the things it lists can be fixed very quickly and normally grant funding or finance options will be available to do this. Soundproofing is a bit more of an investment and a proper installation will require far more short-term disruption than treating a damp problem. I cannot understand why estimated noise in the home is not indicated on every survey. It would help so many people that are now financially trapped in their home with a noise problem that is affecting their health.

An indication of how well neighbour noise, road traffic or environmental noise is dealt with would help so many people. It would encourage better workmanship; it would make many organisations aware of noise in homes. Customers will start demanding the survey and developers will have to start implementing effective sound insulation in homes. A simple traffic light system or noise rating similar to the energy rating would be enough to help buyers with any decision, and at the very least give them an insight to the expected noise in their new home.

To do this, we need to create awareness of the noise in homes, shine a light on the poor building methods used to build modern homes and raise the standards of building insulation for noise.

If you want to reduce unwanted noise in your home, hopefully now you realise it is not just a case of fixing something to the party wall, more than a day's work. The noise in homes will increase, with modern gadgets, flexible employment, the growing population and the busier lives we live. It is inevitable we will all suffer some unwanted noise in our homes. The four-step soundproofing method will identify the noise in your home, and reduce it to an acceptable level – and maybe even so much that you simply don't hear unwanted noise anymore.

If you are a homeowner with a noise problem, you may have attempted soundproofing yourself, or had a builder install soundproofing to the direct noise path alone, or insulation and sound-resistant plasterboard. You may also have given a bathroom or kitchen renovation priority over soundproofing your home, not realising that soundproofing could include these renovation projects (or require them to be pulled out and replaced).

There is no disguising that proper soundproofing is expensive and disruptive (often, more than a fortnight's work).

But soundproofing can change your life!

What Should You Do Next?

The Noise Free Home helps homeowners struggling with noise from neighbours, the considerate noise-maker who wants to avoid bothering their neighbours and anyone who suffers environmental noise such as a busy road. Whatever your decision is, the four-step soundproofing method can help.

Renovating? If you are planning any renovation to your home (extension, making parts open-plan, loft conversion etc), then you are now armed with the right techniques to ensure you don't accidentally create a noise problem for your family or your neighbours. By reading this before starting your renovation, you should be able to make sure your specification prevents unwanted noise from entering your home.

Buying? Construction techniques being used today increase unwanted noise and make it louder in homes. By reading this, you've learnt about building techniques and features that you may want to look out for while home-hunting. Now you are forewarned, you can look for these problems and make informed decisions.

Moving in? Now you know not to put soundproofing on hold until you have fitted a new kitchen, bathroom or flooring. Putting that soundproofing investment on hold and doing another area could be a really expensive mistake. If that other project is connected to a wall or floor that transmits noise, it could be too expensive to remove it later, before installing soundproofing.

If you are a homeowner and have a noise problem, *The Noise Free Home* will help you make informed decisions whether to invest in soundproofing, and give you insights to save you time and money if you do.

If this book has convinced you that you can solve your noise problem once and for all, I would be delighted to hear from you, to come and give you a consultation. Simply visit our website: www.quietco.uk and book a free noise diagnosis, a friendly chat about your noise problem. There is no obligation for a home visit or consultation. We do not do any hard sell; we simply have a passion for helping you with your noise problem. Fill out the website form and get on your way to reducing the noise.

We are a small dynamic team, well known for our effectiveness in reduction in noise, professional service and quality of installation. We can only help a limited amount of people each year.

We leave families feeling good about their home, comfortable and settled, and they don't ever want to move. We will reduce the noise allowing you to the stay in home you love.

Additional Case Studies

Soundproofing a semi-detached house

Simon and Claire loved their area and their street; their house was very practical and had everything they needed. The only problem was the noise from the neighbours.

They could hear children crying, TV, conversations, toilet flushing and when voices were raised, they could even make out the words. It was so bad it was as if they were in the room with them. In the living room and dining room in particular they could hear impact noises, such as plug sockets and light switches and kids running around. In the main bedroom, they could follow conversation and literally had no privacy unless they talked under their breath.

Unfortunately, these noises from the neighbours continued into the night and this made living there very uncomfortable. Due to the noise, Simon and Claire were looking to move to a new house and this made them feel really unsettled.

We carried out the first three steps of the four-step soundproofing method:

Step 1: The main cause was uncovered and treated: A hole in the bedroom separating party wall going all the way through to next door and a poor renovation technique of a hollow stud wall perpendicular to the separating party wall in the kitchen downstairs.

Step 2: Treat the direct noise paths: Our trademarked floating frame wall system was installed to the party wall downstairs and our 3 ½ inch independent wall to the party wall upstairs.

Step 3: Treat the indirect noise paths: A 30mm sound-deadening flanking wall system was installed to all the window walls and the suspended floor void on the ground floors, and all the ceilings were also treated.

And once the soundproofing systems were completed, we implemented the final step:

Step 4: Treat the fixtures and fittings: These were installed by our expert team ensuring the soundproofing was not compromised. Claire and Simon took advantage of the opportunity of a new room to install a large mirror and a new vertical radiator giving them more space for furniture and improved heating distribution in the room.

After soundproofing, Claire and Simon no longer hear their neighbours. In the few weeks after soundproofing, they literally had to go outside and check the neighbours' car was on the drive to see if they were in. Claire described the feeling after soundproofing as 'an absolute dream' and wished she had done it years ago!

Don't just treat the party wall

Sally lived in a semi-detached house in Gloucestershire; she loved the area and knew all the residents in her street. The neighbours were a lovely, busy family but just too noisy.

After researching the noise, she was convinced the noise was coming from the separating party wall alcoves and possibly the chimney stacks. She decided to get the party wall in all rooms of her property soundproofed. Sally got on the internet and decided to purchase a top of the range panel system from a company, which they installed to the alcoves and also all the chimney stacks in the four rooms along the party wall. Although the company had a massive amount of happy customers that gave glowing reviews on a well-known website that checks tradesmen out, there was no change to the noise from the neighbours and all the same noises still came through. Sally was still woken up and felt uncomfortable in her own home.

During our consultation, Sally was shown how the noise was actually entering her house and realised the main cause of the problem had not been addressed and neither had the indirect paths. The soundproofing attempt on the direct path was not adequate for the noises Sally was experiencing from the neighbours and a more substantial system was required.

There were enormous holes below the floor boards, where I could get my arm through and touch next door's floor. All the flanking indirect transmission noise paths were louder than the direct noise paths for her house. As a result, the flanking areas of her house needed to be addressed as well as redoing the soundproofing attempt to deal with the direct noise paths. This meant uninstalling built-in wardrobes again and removing bespoke alcove shelving that Sally had just painted and decorated.

Sally was reluctant to go through all the disruption and upheaval that would come with another house renovation. She had already spent £10,000; how did she know it would really work this time?

Her concerns changed when she was physically shown the mistakes the previous company had made... or what they hadn't done. After educating herself on the noise and also speaking with people that had been through the same decision process as her, Sally decided to invest in her house to try to make it a home once and for all.

We installed our 3 ½ inch, 10-layered system to the direct noise path, which not only sound deadened the wall but also decoupled her new wall from the structure.

Sally does not hear her neighbours anymore. When asked what she heard over the Christmas period she said, 'I think they were away over Christmas as I did not hear anything.' We had independent UKAS-accredited engineers test the party wall before and after to document the results, which is a valuable asset if Sally ever came to sell the property. We managed to speak with the neighbours during the sound testing and they said they had had the whole family round for Christmas and were not away. I remember the amount of bottles in the recycling boxes outside in those first few days of January when we did the after testing; they were definitely *in* over Christmas.

Sally had to remove the 'top of the range' panel system. She then found and treated the main cause of the problem, treated the direct and indirect noise paths, and had her wardrobes, shelves, fireplace, coving and electrical all installed back in a way that did not compromise the soundproofing investment (fixtures and fittings). She no longer gets woken up by noise from next door, doesn't hear her neighbours' conversation and TV and feels comfortable in her own home again.

Glossary

First fix	Term used in the UK building and construction industry. 'First fix comprises all the work needed to take a building from foundation to putting plaster on the internal walls. This includes constructing walls, floors and ceilings, and inserting cables for electrical supply and pipes for water supply.'	https://en.wikipedia.org/wiki/First_fix_and_second_fix

SAP	The government's approved software for calculating the energy rating of your new property is called the Standard Assessment Procedure. Scores range from 1 to 100, with higher scores meaning lower energy consumption and CO_2 levels. Scores over 100 mean that you are net producers of energy.	www.gov.uk/ guidance/standard-assessment-procedure
UKAS	The United Kingdom Accreditation Service is a national accreditation body recognised by the British government to assess the competence of organisations that provide certification, testing, inspection and calibration services.	

Appendix

Terms used in airborne sound-insulation testing

DnT	Standardised level difference	The single-number field measurement of the airborne sound prevented from transmitting through a separating ceiling, wall or floor, expressed in decibels (dB). The higher the decibel figure, the better the sound insulation. If someone is playing the drums in one room (100dB) and I measure 20dB in the adjoining room, the sound reduction is 80dB. The volume of the room together with the reverberation (echo) and amount of absorption in the room are all taken into account as part of the calculation.

DnT,w	Weighted DnT	DnT,w relates to an onsite test which picks up flanking noise, background noise and other elements such as the way the sound insulation was installed. Due to the many variations that can arise on site, this figure will always be between 5dB and 8dB lower than Rw (see below) for a similar product or system.
DnT,w + Ctr	Weighted DnT + spectrum adaptation term	+ Ctr adjusts DnT,w figure by including a low-frequency correction factor. On site, there may be low-frequency background noise such as traffic noise (and other variations). The Ctr figure is always negative so DnT,w + Ctr will always be lower than DnT,w. If you live in Scotland, Ctr is not used and building regulations require DnT,w of 56dB.
Dw		A term used to refer to onsite sound insulation. Simply, the measure of the source noise (eg acoustic drums) minus the measure of the noise of the drums outside the room.
LnT,w	Weighted, normalised impact sound pressure level	The field measurement used to calculate the impact sound insulation of floors on site. If you are a developer, architect or builder renovating or changing the use of a separating floor or ceiling, you need to have an acoustic test to meet Part E of the Building Regulations, and you currently need the figure to be lower than 64dB to meet the required standards.

Rw	Weighted sound reduction index	An acoustic test carried out in a specialist acoustic testing laboratory, with no flanking or background noise. Special instrumentation is used to ensure accuracy and repeatability of the test. The measurement is in decibels and it describes the airborne sound insulation of the building product or system. Online products should quote Rw. Typically, Rw is 5–10dB higher (better) than DnT,w + Ctr.

Percentage Noise Reduction Example

A wall was tested before we started work and sound insulation was found to average 47dB. After sound-proofing, this same wall was tested and achieved a sound insulation value of 62dB. This is a 139.61% reduction in pressure level of sound from next door.

The calculation is as follows:

(A) $62dB = 0.0251785$ Pa (pascals – measure of sound pressure)
(B) $47dB = 0.00447744$ Pa

$$
\begin{aligned}
\text{Change \%} &= 100 \times (A - B)/((A + B)/2) \\
&= 100 \times (0.0251785 - 0.00447744)/ \\
&\quad ((0.0251785 + 0.00447744)/2) \\
&= 100 \times 0.0207/(0.02966/2) \\
&= 100 \times 1.39608 \\
&= 139.61\%
\end{aligned}
$$

THE NOISE FREE HOME

That gives a 139.61% change in pressure level of sound.

10db is a halving of sound as we hear it, so the 15dB reduction represents a significant reduction.

References

Abbott, R, 'What are noise contours?' (2018), www.
bocaairport.com/faq-items/what-are-noise-contours,
accessed 13 October 2020

Allan, S, 'Which materials carry sound waves best?'
(2018), www.sciencing.com/materials-carry-sound-
waves-8342053.html, accessed 13 October 2020

Astroflame Fireseals Ltd, 'Product data sheet: Acoustic & smoke perimeter seals' (2020), www.astroflame.
com/pdf/perimeter-seal-acoustic-smoke.pdf,
accessed 12 October 2020

Britannica, www.britannica.com/science/sound-physics/Circular-and-spherical-waves for how sound waves propagate and move, accessed 15 October 2020

Building Bulletin 93 (BB93), www.gov.uk/government/publications/bb93-acoustic-design-of-schools-performance-standards, accessed 17 December 2020

Celotex, www.celotex.co.uk/resources/faqs, accessed 15 October 2020

Cirrus Research, '6 steps to resolving noise nuisance complaints', www.cirrusresearch.co.uk/library/documents/ebooks/6-steps-resolving-noise-nuisance-complaints.pdf, accessed 14 October 2020

Clark, Dr C, 'Aircraft noise effects on health', Queen Mary University of London (2015), https://assets.publishing.service.gov.uk/government/uploads/system/uploads/attachment_data/file/446311/noise-aircraft-noise-effects-on-health.pdf, accessed 17 December 2020

Collinson, P, '"UK housebuilding revolution": £65,000 prefab homes go into production', *The Guardian*, 2018, www.theguardian.com/society/2018/nov/30/uk-housebuilding-revolution-65000-prefab-homes-go-into-production, accessed November 24 2020

Grimwood, C, 'Complaints about poor sound insulation between dwellings in England and Wales', *Applied Acoustics* 52, 3/4 (1997), 211–23

Jameson, S, 'Noisy neighbours must pay banker £100,000 in compensation because their floors had no carpet, judge rules', *The Telegraph*, 5 March 2018, www.telegraph.co.uk/news/2018/03/05/noisy-neighbours-must-pay-banker-100000-compensation-floors, accessed 16 October 2020

Jensen, HAR, Rasmussen, B and Ekholm, O, 'Neighbour noise annoyance is associated with various mental and physical health symptoms: Results from a nationwide study among individuals living in multi-storey housing', *BMC Public Health* 19, 1508 (2019), https://bmcpublichealth.biomedcentral.com/articles/10.1186/s12889-019-7893-8, accessed 12 October 2020

Kirk, T, 'The use of sonic "anti-loitering" devices is breaching teenagers' human rights', *Independent*, 2017, www.independent.co.uk/news/long_reads/use-sonic-anti-loitering-devices-breaching-teenagers-human-rights-a7880781.html, accessed 9 November 2020

LABC Warranty, www.labcwarranty.co.uk/blog/reduce-noise-in-your-new-build-with-the-help-of-british-gypsum, accessed 13 October 2020

Maschke, C and Niemann, H, 'Health effects of annoyance induced by neighbour noise', *Noise Control Engineering Journal* 55, 3 (2007), 348–56(9)

NHBC Foundation, '40 facts: Homes, housing and house building today' (2018), www.nhbcfoundation.org/wp-content/uploads/2018/09/NF81-WEB.pdf, accessed 13 October 2020

Office of the Deputy Prime Minister, 'Resistance to the Passage of Sound' (2004, updated to 2015), https://assets.publishing.service.gov.uk/government/uploads/system/uploads/attachment_data/file/468870/ADE_LOCKED.pdf, accessed 12 October 2020

Ortleb, EP and Cadice, R 'The speed of sound in other materials' (1993), www.nde-ed.org/EducationResources/HighSchool/Sound/speedinmaterials.htm, accessed 15 October 2020

Rindel, JH, *Sound Insulation in Buildings* (CRC Press, 2018)

Sahu, MK and Singh, L, 'Critical review on types of bricks type 3: AAC block', *International Journal of Mechanical and Production Engineering* 5, 11 (2017)

Scottish Government, 'Housing and sound insulation: Improving existing attached dwellings and designing for conversions', Local Government and Communities Directorate, guidance (2010), www2.gov.scot/resource/doc/217736/0099123.pdf, accessed 12 October 2020

Section 79(1)(g) and (ga) of the Environmental Protection Act 1990, www.environmentlaw.org.uk, accessed 17 December 2020

Shelter, '"Rigged" housebuilding system means eight in ten families cannot afford new home, says Shelter' (2017), https://england.shelter.org.uk/media/press_release/rigged_housebuilding_system_means_eight_in_ten_families_cannot_afford_new_home,_says_shelter, accessed 24 November 2020

Weinhold, D, 'Sick of noise: The health effects of loud neighbours and urban din', Grantham Research Institute on Climate Change and the Environment, Working Paper No. 213, October 2015, www.lse.ac.uk/GranthamInstitute/wp-content/uploads/2015/10/Working-Paper-213-Weinhold.pdf, accessed 12 October 2020

Acknowledgements

First of all I would like to thank my wife Elizabeth, my life partner and soulmate. Thank you for all your patience and guidance and reading through the book many, many times. Without you this book would never have got written. Thank you very much for taking care of the children during the many weekends and evenings it took to write and edit this book.

Thank you very much to Judith Howles for all your support, talking things over, reading and editing, and offering advice and knowledge as the book was being written. Along with Carrie Fleming, Matt Lewis and Kyre Song. Special thanks to Mr SD Kirk MPhil MSc BEng (Hons) MIOA for checking all the technical acoustic jargon, reading and editing and all your support.

I would like to thank Tom Stanley who has been my wing- man for over a decade; Tom has played a massive part in delighting all the clients mentioned in this book and hundreds more during his time with Quietco.

I would like to thank the thousands of clients we work with at Quietco, from whom I have learnt so much and grown as a result. I would like to thank the team at Quietco who deliver excellent service and go out of their way to please every client through their soundproofing journey.

The team at Rethink Press – Lucy McCarraher, Joe Gregory, Anke Ueberberg, and special thanks to Debbie Jenkins, Catriona Hoyle and John Firth for editing and ensuring my words made sense. Additional thanks to Daniel Priestley for mentoring and coaching and encouraging me to write the book.

The Author

 Jim Prior is a successful entre-
preneur with over fifteen
years of experience installing
soundproofing in the UK. He
believes the unwanted noise
in a home is usually caused
by the construction of the
building.

From an early age he was interested in building and
creating new things; developing and showing off his
skills, he would always look to make something that
everyone else would want or be amazed by. When he
was eight years old, he built a fantastic tree house with
his dad, but was disappointed when the rain dripped
through the gaps in the roof, getting his Lego wet! He

decided to solve this problem. He searched his dad's workshop for materials, and collected black bin liners and dog blankets. With a bit of trial and error, he created a waterproof, insulated tree house that he could not only play Lego in all day, but comfortably sleep in, too.

Jim's desire to put materials together to solve a problem or to create something amazing grew from those early experiments. Now, he gets a real buzz out of seeing a customer delighted with the soundproofing he has installed for them, and knowing they can enjoy their home and do the things they love. He gets satisfaction from every aspect: from explaining the concept to the client, understanding the building materials, designing the soundproofing system, to researching and installing gadgets to make the whole experience extraordinary.

Jim's company, Quietco, proudly helps homeowners across the UK reduce unwanted noise in their lives. Quietco's values and underlying principles of quality, integrity, customer service and doing things right first time come out in everything they do. They love to get feedback from clients and are constantly improving. Jim is honoured to solve so many noise problems; the time and effort he has spent over the years to get the right combination of materials in Quietco's systems for different noises and all kinds of situations has been well worth it to change so many people's lives for the better.

He is particularly proud of the four-step soundproofing method, which he strongly believes is the only way to significantly reduce noise and create the home you have always wanted.

- ⊕ www.quietco.uk

- ⬛ www.facebook.com / quietco.uk

- ⬛ www.linkedin.com / company / quietco

- ⬛ www.twitter.com / quietco_uk

- ⓞ www.instagram.com / quietco_uk

9 781781 335116